Teachers Engaged in Research

Inquiry Into Mathematics Classrooms, Grades 9–12

a volume in
Teachers Engaged in Research

Series Editor
Denise S. Mewborn
University of Georgia

Teachers Engaged in Research

Inquiry Into Mathematics Classrooms, Grades 9–12

---

edited by

Laura R. Van Zoest
Western Michigan University

INFORMATION AGE
PUBLISHING

Greenwich, Connecticut • www.infoagepub.com

Library of Congress Cataloging-in-Publication Data

Teachers engaged in research : inquiry into mathematics classrooms, grades 9-12 / edited by Laura R. Van Zoest.
    p. cm. – (Teachers engaged in research)
 Includes bibliographical references.
 ISBN 1-59311-502-4 – ISBN 1-59311-501-6 (pbk.)
 1. Mathematics–Study and teaching (Secondary) 2. Education, Secondary–Research. I. Title: Inquiry into mathematics classrooms, grades 9-12. II. Van Zoest, Laura R. III. Series.
 QA11.2.T424 2006
 510.71'2–dc22

2006007073

*To classroom teachers everywhere who have a vision for their
teaching and engage in research as a way to achieve it.*

ACKNOWLEDGMENTS

The people listed below made important contributions to this book through their reviews of the chapters within it and I am very grateful to them. In addition to reviewing, Shari Stockero served as my editorial assistant during the summer of 2005 and collaborated with me in summarizing the reviews and providing feedback to the authors during that time. I would also like to thank Hope Smith who was instrumental in preparing the chapters for press and provided encouragement throughout the process. Finally, my husband David and son Benjamin, and the families of all the authors, deserve recognition as, invariably, work on this book took place during the participants' personal time.

Doina Apperti	Sandy Madden
Teresa Ballard	Carol Malloy
Brian Bell	Ralph Mason
Mark Bell	Janelle McFeeters
Sandy Blount	Rebecca McGraw
Susan Boone	Diane Moore
Shelley Bowes	James Muchmore
Laurinda Brown	Cynthia Nicol
Mark Ellis	Blake Peterson
Doug Franks	Amy Ruff
Charles Funkhouser	Jesse Solomon
Julie Gainsburg	Shari Stockero
Christian Hirsch	Sally Torrillo
Sherry Hix	Gwen Zimmerman
Michael Lutz	

LIST OF CONTRIBUTORS

Apolinário Barros	Boston International High School, Boston, MA
Darlene Bazcuk	Colonsay School, Colonsay Saskatchewan
John Carter	Adlai E. Stevenson High School, Lincolnshire, IL
Sharon Christensen	Mountain Ridge Junior High School, Highland, UT
Marilyn Cochran-Smith	Boston College, Newton, MA
Robert Gammelgaard	Adlai E. Stevenson High School, Lincolnshire, IL
Nicole Garcia	Washtenaw Technical Middle College, Ann Arbor, MI
Florence Glanfield	University of Saskatchewan, Saskatoon Saskatchewan
Maureen Grant	North Central High School, Indianapolis, IN
William J. Harrington	The Pennsylvania State University, State College, PA
Scott Hendrickson	Brigham Young University, Provo, UT
Patricio G. Herbst	University of Michigan, Lansing, MI
Craig Huhn	Holt High Schools, Holt, MI
Kellie Huhn	Holt High Schools, Holt, MI
Peg Lamb	Director of NSF Bridges Project, Holt Public Schools, Holt, MI
Vicki Lyons	Lone Peak High School, Highland, UT
P. Janelle McFeetors	River East Collegiate, Winnipeg, Manitoba
Rebecca McGraw	University of Arizona, Tuscan, AZ
Adrianne Olson	Lone Peak High School, Highland, UT
Ann Oviatt	Deslisle Composite High School, Delisle Saskatchewan
Michelle Pope	Adlai E. Stevenson High School, Lincolnshire, IL
Beth Ritsema	Western Michigan University, Kalamazoo, MI
Heather J. Robinson	Grayson High School, Loganville, GA
Dorina Sackman	Stonewall Jackson Middle School, Orlando, FL
Jesse Solomon	Boston Teacher Residency, Boston Public Schools, Boston, MA
Laura R. Van Zoest	Western Michigan University, Kalamazoo, MI
Michael Verkaik	Holland Christian High School, Holland, MI
Gwendolyn Zimmermann	Hinsdale Central High School, Hinsdale, IL

CONTENTS

SERIES FOREWORD

Marilyn Cochran-Smith
Boston College

This series, *Teachers Engaged in Research: Inquiry Into Mathematics Classrooms,* represents a remarkable accomplishment. In four books, one each devoted to teaching and learning mathematics at different grade level groupings (Pre-K–2, 3–5, 6–8, and 9–12), ninety-some authors and co-authors write about their work as professional mathematics educators. Across grade levels, topics, professional development contexts, schools, school districts, and even nations, the chapters in these four books attest to the enormous complexity of teaching mathematics well and to the power of inquiry as a way of understanding and managing that complexity.

In a certain sense, of course, the education community already knows that doing a good job of teaching mathematics is demanding—and all too infrequent—and that it requires deep and multiple layers of knowledge about subject matter, learners, pedagogy, and contexts. Nearly two decades of ground-breaking research in cognition and mathematics education has told us a great deal about this. But the books in this series are quite different from most of what has come before. These four books tell us about teaching and learning mathematics from the inside—from the perspectives of school-based teacher researchers who have carefully studied the commonplaces of mathematics teaching and learning, such as the whole

Teachers Engaged in Research
Inquiry Into Mathematics Classrooms, Grades 9–12, pages ix–xx
Copyright © 2006 by Information Age Publishing
All rights of reproduction in any form reserved.

class lesson, the small group activity, the math problem, the worksheet where the student shows his or her work, the class discussion about solutions and answers, and the teacher development activity. Taking these and other commonplaces of mathematics teaching and learning as sites for inquiry, the multiple authors of the chapters in these four books offer richly textured and refreshingly insightful insider accounts of mathematics teaching and learning. Reflecting on the contrasts that sometimes exist between teachers' intentions and the realities of classroom life, the chapters depict teachers in the process of considering and reconsidering the strategies and materials they use. Drawing on students' work and classroom discourse, the chapters show us what it looks like as teachers strive to make sense of and capitalize on their students' reasoning processes, even when they don't end up with traditional right answers. Staying close to the data of practice, the chapters raise old and new questions about how students—and teachers—learn to think and work mathematically. In short, this remarkable quartet of books reveals what it really means—over time and in the context of different classrooms and schools—for mathematics teachers to engage in inquiry. The goal of all this inquiry is nothing short of a culture shift in the teaching of mathematics—the creation of classroom learning environments where the focus is deep understanding of mathematical concepts, practical application of skills, and problem solving.

These four books provide much-needed and rich detail from the inside about the particulars of mathematics teaching and learning across a remarkably broad range of contexts, grade levels, and curricular areas. In addition, the books tell us a great deal about teachers' and pupils' learning over time (and the reciprocal relationships of the two) as well as the social, intellectual and organizational contexts that support their learning. Undoubtedly teachers and teacher educators will find these books illuminating and will readily see how the authors' successes and the struggles resonate with their own. If this were the only contribution this series made, it would be an important and worthwhile effort. But this series does much more. The books in this series also make an important contribution to the broader education community. Taken together as a whole, the chapters in these four books have the potential to inform mathematics education research, practice and policy in ways that reach far beyond the walls of the classrooms where the work was originally done.

TAKING AN INQUIRY STANCE

This series of books shows brilliantly and in rich and vivid detail what it means for teachers to take an inquiry stance on teaching and learning mathematics in K-12 classrooms over the professional lifespan. "Inquiry stance" is a concept that my colleague, Susan Lytle, and I have written about over the last decade and a half. It grew out of the dialectic of our simultaneous work as teacher educators involved in day-to-day, year-to-year participation in teacher learning communities, on the one hand, and as researchers engaged in theorizing the relationships of inquiry, knowledge, and practice, on the other. In our descriptions of inquiry as stance, we have repeatedly emphasized its distinction from "inquiry as project." By inquiry project (as opposed to stance), we refer to things like the classroom study that is the required culminating activity during the student teaching semester in a preservice program where inquiry is not integral and not infused throughout, or, the day-long workshop on teacher research or action research that is part of a catalog of professional development options for experienced teachers rather than a coherent part of teacher learning over the lifespan.

Teacher research or inquiry *projects* are discrete and bounded activities, often carried out on a one-time or occasional basis. In contrast, as we have suggested (Cochran-Smith & Lytle, 1999), when inquiry is a *stance*, it extends across the entire professional lifespan, and it represents a world view or a way of knowing about teaching and learning rather than a professional project:

> In everyday language, "stance" is used to describe body postures, particularly with regard to the position of the feet, as in sports or dance, and also to describe political positions, particularly their consistency (or the lack thereof) over time. In the discourse of qualitative research, "stance" is used to make visible and problematic the various perspectives through which researchers frame their questions, observations, and interpretations of data. In our work, we offer the term *inquiry as stance* to describe the positions teachers and others who work together in inquiry communities take toward knowledge and its relationships to practice. We use the metaphor of stance to suggest both orientational and positional ideas, to carry allusions to the physical placing of the body as well as to intellectual activities and perspectives over time. In this sense, the metaphor is intended to capture the ways we stand, the ways we see, and the lenses we see through. Teaching is a complex activity that occurs within webs of social, historical, cultural and political significance. Across the life span an inquiry stance provides a kind of grounding within the changing cultures of school reform and competing political agendas.
>
> *Inquiry as stance* is distinct from the more common notion of inquiry as time-bounded project or activity within a teacher education course or professional development workshop. Taking an inquiry stance

means teachers and student teachers working within inquiry communities to generate local knowledge, envision and theorize their practice, and interpret and interrogate the theory and research of others. Fundamental to this notion is the idea that the work of inquiry communities is both social and political—that is, it involves making problematic the current arrangements of schooling, the ways knowledge is constructed, evaluated, and used, and teachers' individual and collective roles in bringing about change. To use *inquiry as stance* as a construct for understanding teacher learning in communities, we believe that we need a richer conception of knowledge than that allowed by the traditional formal knowledge-practical knowledge distinction, a richer conception of practice than that suggested in the aphorism that practice is practical, a richer conception of learning across the professional life span than that implied by concepts of expertise that differentiate expert teachers from novices, and a rich conception of the cultures of communities as connected to larger educational purposes and contexts . (pp. 288–289)

Developing and sustaining an inquiry stance on teaching, learning and schooling is a life-long and constant pursuit for beginning teachers, experienced teachers, and teacher educators alike. A central aspect of taking an inquiry stance is recognizing that learning to teach is a process that is never completed but is instead an ongoing endeavor. The bottom-line purpose of taking an inquiry stance, however, is not teachers' development for its own sake (although this is an important and valuable goal), but teachers' learning in the service of enriched learning opportunities for Pre-K–12 students.

The chapters in these four books vividly explicate the work of Pre-K–12 teacher researchers who engage in inquiry to transform their teaching practice, expand the mathematical knowledge and skill of their students, and ultimately to enhance their students' life chances in the world. Across the 50 chapters in these books, the serious intellectual work of mathematics teaching is revealed again and again as are the passion teachers have for their work (and their students) and the complexity of the knowledge teachers must have to support students' emerging mathematical development.

INQUIRY INTO TEACHING AND
LEARNING IN MATHEMATICS

Over the last two decades, various forms of practitioner inquiry, such as teacher research, action research, and collaborative inquiry, have become commonplace in preservice teacher preparation programs as well as in professional development projects and school reform efforts of various kinds. Although there are many variations and often different underlying assumptions among these, the use of inquiry has generally reflected a move away from transmission models of teacher training and retraining wherein teachers are expected to implement specific practices developed by others. The move has been toward a concept of teacher learning as a life-long process of both posing and answering questions appropriate to local contexts, working within learning communities to construct and solve problems, and making all of the aspects of the work of teaching possible sites for inquiry.

As this series of books demonstrates so clearly, an inquiry stances is quite compatible with the professional standards for teaching and learning mathematics that have emerged since the mid 1980s. In fact, in a number of the chapters in these books, the explicit purpose of the researchers is to study the implementation and effectiveness of standards-driven mathematics teaching and learning. The chapters concentrate on a wide range of mathematics topics, for example: concepts of number, addition and subtraction, algebraic thinking, linear measurement, geometric patterns and shapes, classification and sorting, mathematical proofs, multiplicative reasoning, division of fractions, data analysis and probability, volume, and the concept of limit. The chapters also examine what happens in terms of students' learning and classroom culture when various strategies and teaching methods are introduced: bringing in new models and representations of mathematical concepts and operations, teaching with problems, using writing in mathematics instruction, having students share their solution strategies, supporting students' own development of the algorithms for various mathematical operations, encouraging students to work in small groups and other collaborative arrangements, including role play in one's repertoire of mathematics teaching strategies, and integrating multiple opportunities for students to participate in hands-on technology and to work with technology-rich problems. Taken as a group, the chapters in these four volumes illustrate the power of an inquiry stance to widen and deepen teachers' knowledge of the subject matter of mathematics at the same time that this stance also enhances teachers' understanding of the learning processes of their students. Both of these—deeper knowledge of mathematics and richer understandings of learners and learning—are essential ingredients in teaching mathematics consistent with today's high standards.

Some of the chapters in these four volumes are written by single authors. These chronicle the ongoing efforts of teachers to understand what is going on in their classrooms so they can build on the knowledge and experiences students bring at the same time that they expand students' knowledge and skill. It is not at all surprising, however, that many of the chapters are co-authored, and even the single-authored chapters often reflect teachers' experiences as part of larger learning communities. Inquiry and community go hand in hand. The inquiries reported in these volumes feature collaborations among teachers and a whole array of colleagues, including fellow teachers, teacher study groups, teacher educators, university-based researchers, professional development facilitators, curriculum materials developers, research and development center researchers, and directors of large-scale professional development projects.

Across the chapters, there is a fascinating range of collaborative relationships between and among the teacher researchers and the subjects/objects of their inquiries: two teachers who collaborate to expand their own knowledge of mathematical content, the teacher researcher who is the subject of another researcher's study, the school-based teacher researcher and the university-based researcher who form a research partnership, the teacher who collaborates with a former teacher currently engaged in graduate study, pairs and small groups of teachers who work together to implement school-wide change in how mathematics is taught, teachers who team up to examine whether and how standards for mathematics teaching and learning are being implemented in their classrooms, teachers who engage in inquiry as part of their participation in large-scale professional development projects, the teacher who is part of a team that developed a set of professional development materials, a pair of teachers who collaborate with a university-based research and development group, a group of teachers who develop an inquiry process as a way to induct new teachers into their school, two teachers who inquire together about pedagogy and practice in one teacher's classroom, and teachers who form research groups and partnerships to work with teacher educators and student teachers. Almost by definition, practitioner inquiry is a collegial process that both occurs within, and stems from, the collaborations of learning communities. What all of these collaborations have in common is the assumption that teacher learning is an across-the-professional-lifespan process that is never "finished" even though teachers have years of experience. The chapters make clear that learning from and about teaching through inquiry is important for beginning and experienced teachers, and the intellectual work required to teach mathematics well is ongoing.

The inquiries that are described in these four books attest loudly and clearly to the power of questions in teachers' and students' learning. These inquiries defy the norm that is common to some schools where the competent teacher is assumed to be self-sufficient and certain and where asking questions is considered inappropriate for all but the most inexperienced teachers. Similarly, the chapters in these four books challenge the myth that good teachers rarely have questions they cannot answer about their own practices or about their students' learning. To the contrary, although the teachers in these four books are without question competent and may indeed be self-sufficient and sometimes certain, they are remarkable for their questions. Teachers who are researchers continuously pose problems, identify discrepancies between theory and practice, and challenge common routines. They continuously ask critical questions about teaching and learning and they do not flinch from self-critical reflection: Are the students really understanding what the teacher is teaching? What is the right move to make at various points in time to foster students' learning? How does the teacher know this? Are new curricular materials and teaching strategies actually supporting students' learning? How do students' mathematical ideas change over time? What is the evidence of students' growth and development? How can theory guide practice? How can practice add to, even alter, theory? These teachers ask questions not because they are failing, but because they—and their students—are learning. They count on their collaborators for alternative perspectives about their work and alternative interpretations of what is going on. In researching and writing about their work, they make explicit and visible to others both the decisions that they make on an ongoing basis and the intellectual processes that are the backdrop for those decisions. Going public with questions, seeking help from colleagues, and opening up one's practice to the scrutiny of outsiders may well go against the norms of appropriate teaching behaviors in some schools and school districts. But, as the chapters in this series of books make exquisitely clear, these are the very activities that lead to enriched learning opportunities and expanded knowledge for both students and teachers.

In this current era of accountability, there is heavy emphasis on evidence-based education and great faith in the power of data and research to improve educational practice. The chapters in the four books in this series illustrate what it looks like when mathematics teaching and learning are informed by research and evidence. In some of the chapters, for example, teacher researchers explicitly examine what happens when they attempt to implement research-based theories and teaching strategies into classroom practice. In these instances, as in others, practice is driven by research and evidence in several ways. Teacher researchers study the work of other researchers, treating this work as generative and illuminating, rather than regarding it as prescriptive and limiting. They reflect continuously about how others' research can (and should) inform their curricular, instructional, and assessment decisions about teaching mathematics. They also examine what consequences these decisions have for students' learning by collecting and analyzing a wide array of classroom data—from young children's visual representations of mathematical concepts and operations to classroom discussions about students' differing solutions to math problems to pre- and post-tests of students' mathematical knowledge and skill to interviews with students about their understandings to students' scores on standardized achievement tests.

In many of the chapters, teachers explicitly ask what it is that students know, what evidence there is that students have this knowledge and that it is growing and developing, and how this evidence can be used to guide their decisions about what to do next, with whom, and under what conditions. Braided together with these lines of inquiry about students' learning are questions about teachers' own ways of thinking about mathematics knowledge and skill. It is interesting that in many of the inquiries described in these four books, it is difficult to separate process from product or to sort out instruction from assessment. In fact, many of the chapters reveal that the distinctions often made between instruction and assessment and between process and product are false dichotomies. When mathematics teaching is guided by an inquiry stance and when teachers make decisions based on the data of practice, assessment of students' knowledge and skill is ongoing and is embedded into instruction, and the products of students' learning are inseparable from their learning and reasoning processes.

Finally, in some of the chapters in this quartet of books, teacher researchers and their colleagues use the processes of inquiry to examine issues of equity and social justice in mathematics teaching and learning. For example, one teacher problematizes commonly used phrases and ideas in mathematics education such as "success," particularly for those who have been previously unsuccessful in mathematics. Another explores connections between culture and mathematics by interviewing students. A trio of teachers examines the mathematical understandings of their students with disabilities, while another teacher chronicles his efforts to construct mathematics problems and projects that focus on social justice. Each of these educators values and draws heavily on the data of students' own voices and perspectives; each works to empower students as active agents in their own learning. These examples are a very important part of the collection of inquiries in this book series. When scholars write about teaching and teacher education for social justice, mathematics is the subject matter area least often included in the discussion. When teachers share examples of teaching for social justice in their own schools and classrooms, mathematics is often the area they have the most difficulty incorporating. When student teachers plan lessons and units related to equity and diversity, mathematics is often the subject area for which they cannot imagine any connection. The chapters in these books that specifically focus on equity and social justice make it clear that these issues readily apply to mathematics teaching and learning. But even in the many chapters for which these issues are not the explicit focus, it is clear that the teachers' intention is to empower all students—even (and especially) those groups least well served by the current educational system—with greater mathematical acuity and agency.

THE VALUE OF TEACHERS' INQUIRIES INTO MATHEMATICS TEACHING AND LEARNING

There is no question that engaging in inquiry about teaching and learning mathematics is an important (and often an extremely powerful) form of professional development that enhances teachers' knowledge, skill and understandings. This is clear in chapter after chapter in the four volumes of this series. The authors themselves describe the process of engaging in self-critical systematic inquiry as transformative and professionally life-changing. They persuasively document how their classroom practices and their ways of thinking about mathematical knowing changed over time. They combine rich and multiple data sources to demonstrate growth in their students' knowledge and skill. Undoubtedly many Pre-K–12 teachers and school- and university-based teacher educators will find the inquiries collected in these four volumes extraordinarily helpful in furthering their own work. The questions raised by the teacher researchers (often in collaboration with university-based colleagues and others) will resonate deeply with the questions and issues of other educators striving to teach mathematics well by fostering conceptual understandings of big mathematical ideas, reliable but imaginative problem solving strategies, and solid mathematical know-how about practical applications to everyday problems. The teacher researchers whose work is represented in these volumes have made their questions and uncertainties about mathematics teaching explicit and public. In doing so, they have offered their own learning as grist for the learning and development of other teachers, teacher educators, and researchers. This is a major contribution of the series.

Few readers of the volumes in this series will debate the conclusion that systematically researching one's own work as a mathematics educator is a valuable activity for teacher researchers themselves, for their students in Pre-K–12 schools and classrooms, and for other Pre-K–12 teachers and teacher educators who are interested in the same issues. But some readers will raise questions about whether or not there is value to this kind of inquiry that carries beyond the participants involved in the local context or that extends outside the Pre-K–12 teaching/teacher education community. After all, the argument of some skeptics might go, for the most part the inquiries about mathematics teaching and learning that are included in this series were prompted by the questions of individual teachers or by small local groups of practitioners working in collaboration. With a few exceptions, these inquiries were conducted in the context of a single classroom, course, school, or program. Almost by definition then, the skeptics might say, few of the inquiries in these four volumes can hold up to traditional research criteria for transferability and application of findings to other populations and contexts (especially if conceptualized as the identification of causes and effects). For this reason, the skeptics might conclude, the kind of work included in this series is nothing more than good professional development for individuals in local communities, which may be of interest to other teacher educators and professional developers.

Putting aside for the moment that good professional development is essential in mathematics education (not to mention, hard to come by), the skeptics' line of argument, as outlined in the paragraph above, is both short-sighted and uninformed. The inquiries that are part of the *Teachers Engaged in Research: Inquiry Into Mathematics Classrooms* series are valuable and valid well beyond the borders of the local communities where the work was done, and the criteria for evaluating traditional research are not so appropriate here. With many forms of practitioner inquiry, appropriate conceptions of value and validity are more akin to the idea of "trustworthiness" that has been forwarded by some scholars as a way to evaluate the results of qualitative research. From Mishler's (1990) perspective, for example, the concept of validation ought to replace the notion of validity. Validation is the extent to which a particular research community, which works from "tacit understandings of actual, situated practices of a field of inquiry" (Lyons and LaBoskey, 2002, p. 19), can rely on the concepts, methods, and inferences of an inquiry for their own theoretical and conceptual work. Following this line of reasoning, validation rests on concrete examples (or "exemplars") of actual practices presented in enough detail so that the relevant community of practitioner researchers can judge the trustworthiness and usefulness of the observations and analyses of an inquiry.

Many of the inquiries in these four books offer what may be thought of as exemplars of teaching and learning mathematics that are consistent with current professional standards in mathematics education. Part of what distinguishes the inquiries of the teacher researchers in these four books from those of outside researchers who rely on similar forms of data collection is that in addition to documenting students' learning, teacher researchers have the opportunity and the access to systematically document their own teaching and learning as well. They can document their own thinking, planning, and evaluation processes as well as their questions, interpretive frameworks, changes in views over time, issues they see as dilemmas, and themes that recur. Teacher researchers are able to analyze the links as well as the disconnects between teaching and learning that are not readily accessible to researchers based outside the contexts of practice. Systematic examination and analysis of students' learning juxtaposed and interwoven with systematic examination of the practitioners' intentions, reactions, decisions, and interpretations make for incredibly detailed and complex analyses—or exemplars—of mathematics teaching and learning.

With rich examples, the teacher researchers in these books demonstrate how they came to understand their students' reasoning processes and thus learned to intervene more adeptly with the right question, the right comment, a new problem, or silent acknowledgement and support. Braiding insightful reflections on pedagogy with perceptive analyses of their students' understandings, the authors in these four books make explicit and visible the kinds of thinking and decision making that are usually implicit and invisible in studies of mathematics teaching. Although they are usually invisible, however, these ways of thinking and making teaching decisions are essential ingredients for teaching mathematics to today's high standards. I believe that a critical contribution of this series of books—a contribution that extends well beyond the obvious benefits to the participants themselves—is the set of exemplars that cuts across grade levels and mathematics topics. What makes this series unique is that the exemplars in this book do not emerge from the work of university-based researchers who have used the classroom as a site for research and for the demonstration of theory-based pedagogy. Rather the trustworthiness—or validation—of the exemplars in these four books derives from the fact that they were conducted by school-based teacher researchers who shouldered the full responsibilities of classroom teaching while also striving to construct appropriate curriculum, develop rich teaching problems and strategies, and theorize their practice by connecting it to larger philosophical, curricular, and pedagogical ideas.

This series of books makes a remarkable contribution to what we know about mathematics teaching and learning and about the processes of learning to teach over time. Readers will be thoroughly engaged.

REFERENCES

Cochran-Smith, M., & Lytle, S. L. (1999). Relationship of knowledge and practice: Teacher learning in communities. In A. Iran-Nejad & C. Pearson (Eds.), *Review of research in education* (Vol. 24, pp. 249–306). Washington, DC: American Educational Research Association.

Lyons, N., & LaBoskey, V. K. (Eds.). (2002). *Narrative inquiry in practice: Advancing the knowledge of teaching.* New York: Teachers College Press.

Mishler, E. (1990). Validation in inquiry-guided research: The role of exemplars in narrative studies. *Harvard Educational Review, 60*(4), 415–442.

CHAPTER 1

INTRODUCTION TO THE 9–12 VOLUME

Laura R. Van Zoest[1]
Western Michigan University

What is a teacher's role in research? Certainly teachers have been the *impetus* of research studies throughout the history of educational research. They also have been referred to as *consumers* of research. It is unusual, however, for teachers to be considered *producers* of research. Yet it is not unusual for teachers to be engaged in inquiry. In fact, some would consider this a fair description of what it means to teach. The difference is that, for most teachers, the process as well as the product of their inquiry is tacit. It may not be well defined in terms of specific research questions or systematic in terms of data collection and analysis. Furthermore, their work may not be presented to colleagues for discussion and review or disseminated for publication. Those scholarly activities have typically not been part of the culture of teaching in most school systems in North America.

That may be changing. The chapters in this volume provide examples of the ways in which 9–12 grade mathematics teachers from across North America are engaging in research. The authors who contributed to this volume were involved in research through a variety of activities, including:

- reading and reflecting on a variety of research and other literature in the field;

Teachers Engaged in Research
Inquiry Into Mathematics Classrooms, Grades 9–12, pages 1–18
Copyright © 2006 by Information Age Publishing

- interpreting findings from the research literature to influence their instructional practice;
- participating in study groups with their colleagues;
- generating research questions for themselves and others to investigate;
- participating in research studies and professional development projects led by other researchers; and
- designing and implementing their own studies and sharing their findings.

It is clear from these chapters that the authors' involvement in these research activities had a profound impact on their conceptions of teaching and learning. Their teaching changed as a result of their inquiries and thus these research activities had an immediate, direct impact on practice. Through the sharing of their stories we hope to broaden that impact beyond their classrooms and schools to the wider mathematics education community.

These chapters provide us a glimpse of the questions that capture the attention of teachers, the methodologies that they use to gather data, and the ways in which they make sense of what they find. Some of the research findings are preliminary, tentative; others are confirmatory; and some are groundbreaking. In all cases, they provide fodder for further thinking and discussion about critical aspects of mathematics education.

HOW THIS BOOK CAME TO BE

History of the Series

The idea for the Teachers Engaged in Research series arose from efforts by the Research Advisory Committee (RAC) of the National Council of Teachers of Mathematics (NCTM) to expand traditional conceptions of research in mathematics to include practitioner inquiry and research questions that are of interest to practitioners. Beginning in the late 1990's, the RAC made a concerted effort to recognize teachers as producers (and not just consumers) of research. To this end, NCTM sponsored a Working Conference on Teacher Research in Mathematics Education in Albuquerque, NM in 2001. The goal of this conference was to articulate a list of issues that should be considered in developing a framework for teacher research in mathematics education. This conference led to a grant proposal that would have brought together participants from the conference for a writing workshop that was to lead to a publication similar to this one. For a variety of reasons, that project did not come to fruition. Around 2000 the RAC turned its attention to including teacher researchers as both attendees and speakers at the Research Presession that precedes the NCTM annual meeting.

During this time there was ongoing dialogue in the RAC about a publication that would highlight practitioner inquiry. Simultaneously, the RAC and the Educational Materials Committee (EMC) decided that it was time to issue an update of the *Handbook of Research on Mathematics Teaching and Learning* (Grouws, 1992). Unbeknownst to many, the series titled *Research into Practice* edited by Sigrid Wagner was originally conceived by the EMC as a "companion" to the *Handbook*. The *Research into Practice* series featured chapters co-authored by university-based researchers and classroom teachers and reflected a view of teachers as consumers of research—an accurate reflection of the field at the time. When the RAC and EMC discussed the update of the *Handbook* and its companion, the companion was recast to reflect a view of teachers as producers of research. The *Handbook* (Lester, in preparation) documents a portion of the knowledge base available in mathematics education; the four volumes of the *Teachers Engaged in Research* series fill a gap in that knowledge base by focusing on the research contributions of classroom teachers, thus allowing a different set of voices to be heard.

Purpose of the Series

The goal of this series is to use teachers' accounts of classroom inquiry to make public and explicit the processes of doing research in classrooms. Teaching is a complex, multi-faceted task, and this complexity often is not captured in research articles. Our goal is to illuminate this complexity. Research that is done in classrooms by and with teachers is necessarily messy, and our stance is that the ways in which this is so should be articulated, not hidden.

Identifying Authors

Using the participants of the Albuquerque conference as a starting point, the editorial board for the series generated a list of teachers, projects, and university faculty who we knew had been engaged in classroom research themselves or might know of others who had. Through personal contacts with these individuals, and those they led us to, we compiled a list of potential authors, the nature of their work, and relevant background information. For each volume, we attempted to generate a set of potential authors and topics that would span aspects of the *Principles and Standards for School Mathematics* (NCTM, 2000), such as the content standards, process standards, and principles. We also tried to find a set of pieces that would span various roles that teachers might play in research—principal researcher, co-researcher, research participant, and consumer of research. We then invited authors for specific chapters in accordance with the categories outlined above. Authors were specifically asked to highlight the following aspects of their work in their manuscripts:

- mathematical content and processes that were addressed (using the *Principles and Standards* document as a guide)
- demographic data on school/student population/community (as it pertains to the research)
- the authors' role in the research and what the experience was like for them
- data sources for the work (incorporated into the narrative as appropriate, e.g,. interview transcripts, student work, teacher reflections, summary of a class session)
- explanation of the data analysis process (How did the authors make sense of their data?)
- articles that influenced their work (rather than a full literature review, references to work that may have sparked their curiosity, helped them think of data collection methods, contrasted their findings, or was helpful in another way)
- implications (How did this influence the authors' subsequent teaching? What might others take away from this study?)

We particularly stressed the need for the manuscripts to reflect disciplined inquiry and for claims to be based on evidence.

The Review Process

The editorial board took great care with the review process in an effort to be respectful of the varied experiences the authors had with writing for publication. Initial outlines and drafts were reviewed and commented on by the volume editor to ensure that the general flavor of the manuscript met the goals for the series. For the 9–12 volume, more polished drafts were reviewed by three or more reviewers—including at least one teacher and one university teacher educator. To the extent possible, we solicited reviewers who had experience with teacher inquiry. We asked reviewers to be sensitive to the fact that the authors were making themselves vulnerable by sharing their research with a wider audience, and we reminded them that the type of research reflected in the manuscripts is much messier than research that is traditionally reported in journal articles. Thus, we asked the reviewers to approach their reviews from a mentoring frame of mind, striving to both support the authors and strengthen the manuscript. Feedback from reviewers was not given to authors verbatim; instead it was compiled by the editors to ensure that the authors received a consistent and helpful message from the review process.

How These Volumes Might Be Used

This series was written primarily for teachers who have or who are being encouraged to develop an awareness of and commitment to teaching for understanding. The research findings presented in these chapters suggest instructional implications worthy of teachers' consideration. Often, authors have described instructional practices or raised issues that have the potential to broaden views of teaching and learning mathematics. Embedded in the chapters are interesting problems and tasks used in the authors' work that teachers could use in their own classrooms. A hallmark of good research is its connection to the existing literature in the field, and the authors of this volume have drawn from the research literature to inform their work. The reference lists accompanying each chapter can be useful resources and should not be overlooked. In addition, this series showcases the variety of ways teachers can become engaged in research and it is hoped that readers will recognize that teacher research can be accessible and potentially beneficial to their own work in their classroom. The volumes could be particularly useful for teacher study groups, courses for preservice and inservice teachers, and professional development projects.

This is not to imply, however, that the series is intended only for teachers. The volumes are interesting, informative resources for other researchers, policy makers, school administrators, and teacher educators as well. In particular, they provide an opportunity for those outside the classroom to gain insight into the kind of issues that matter to teachers, the ways in which those issues might be researched, and the contributions that research can make to the work of classroom teachers.

OVERVIEW OF CHAPTERS AND COMMON THEMES

Table 1.1 provides an overview of the chapters in this volume. There are many ways in which the chapters could be categorized; the columns in Table 1.1 represent four in which there were both differences and similarities across the thirteen chapters. The categories—Research Type, Focus of Chapter, Methodology, and Research Issues—are discussed in the sections below.

Table 1.1.   Overview of Teachers Engaged in Research:
            Inquiry Into Mathematics Classrooms, Grades 9–12

Author	Research Type	Focus of Chapter	Methodology	Research Issue
Zimmermann	Focused Study	Probability Simulation	Teaching Experiment	Research Witness
Harrington	Focused Study	Limit Concept/ Computing Technology	Student Assessments & Interviews	Extending Research of Others
Robinson	Practical Inquiry	Instructional Change	External Assessment	Applying Research
Glanfield, Baczuk, & Oviatt	Practical Inquiry	Mathematical Communica- tion	Analyzing Student Work & Presentations	Classroom Teacher/ University Researcher Collaboration
Huhn, Huhn & Lamb	Practical Inquiry	Student Alge- braic Thinking	Interview Problems	Professional Conversations
Carter, Gammelgaard & Pope	Practical Inquiry	Lesson Study/ Teacher Induction	Analyzing Lesson	Applying Research
Hendrickson & Christensen	Practical Inquiry	Extending Ele- mentary Level Research to Secondary Level	Analyzing Stu- dent Work	Extending Research of Others
McFeetors	Social Inquiry	Nature of Success	Listening	Theory Building
Solomon	Social Inquiry	Culture and Pedagogy	Student Interviews	Examining Underlying Assumptions
Garcia & Herbst	Part of Larger Project	Teaching Mathematics with Prob- lems/Fairness	Classroom Data & Student Interviews	Research as Self-Reflection
Barros & Sackman	Part of Larger Project	Instructional Technology/ Mathematics of Motion	Analyzing Classroom Interaction	Formal Research Col- laboration

Table 1.1. Overview of Teachers Engaged in Research:
Inquiry Into Mathematics Classrooms, Grades 9–12

Author	Research Type	Focus of Chapter	Methodology	Research Issue
Grant & McGraw	Part of Larger Project	Classroom Discourse	Analyzing Classroom Interaction	Classroom Teacher/ University Researcher Collaboration
Verkaik & Ritsema	Part of Larger Project	Instructional Change/ Reform Curricula	Analyzing Teaching	Research as Professional Development

Research Type and Focus of Chapter

Although there is some overlap, the studies can be divided into four distinct research types: *Focused Study, Practical Inquiry, Social Inquiry,* and *Part of a Larger Project.* Two of the chapters, Gwendolyn Zimmermann's and William Harrington's can be given the label *Focused Study.* Even though these chapters differed in many ways, they shared a focus on learning about students' thinking in a specific content area—Zimmermann in the area of probability simulation and Harrington limit concept development in a technological environment. These studies were also tightly defined in that they used clearly articulated instruments to collect their data and built on a traditional review of the literature. Harrington's chapter, in particular, raises the question of what makes this research different from that which is done by university researchers. Does it matter that the researcher was also a teacher? What about teachers who set aside their teacher hat to return to the university for an advanced degree? At what point do they cease to be a teacher-researcher and become a university researcher? What if they return to the classroom with all the skills they learned at the university? Is it a frame of mind or a person's position that determines what kind of researcher they are? One thing that is clear in Harrington's chapter, and throughout the book, is that the researcher's role as a classroom teacher affects the types of research questions that he or she asks and the way in which those answers are sought.

The label *Practical Inquiry* represents five studies whose authors had a particular question about their teaching, or that of their department or district, to which they wanted an answer. In many ways, the work reported on in these chapters is consistent with what is often called Action Research. The researchers learned about the work of others, tried it out in their settings, and assessed the results. Another commonality among the teachers in this category is that they engaged in research specifically to help them respond to changes called for by the NCTM *Standards*. This comes across in a variety of ways. For example, Heather Robinson used research to inform the changes she made in her classroom instruction and to assess whether those changes had the anticipated effect on her students' learning and performance on an existing assessment. Florence Glanfield, Ann Oviatt, and Darlene Baczuk looked for evidence in students' work and interactions that curricular and pedagogical changes they had made to support mathematical communication were having the desired effects. Craig Huhn, Kellie Huhn, and Peg Lamb interviewed students with disabilites to gain insight into their algebraic thinking as part of efforts to support mathematics learning for all students.

The chapters by John Carter, Robert Gammelgaard, and Michelle Pope, and by Scott Hendrickson and Sharon Christensen reflect groups of teachers who were working together in an extended way to improve the teaching in their school systems. Carter did this by implementing a Lesson Study process for inducting teachers new to his high school into an existing culture reflective of the Standards. Henrickson & Christensen took what they learned about research on elementary school children's mathematical thinking and extended it to the secondary school level as a way to prepare for a district-wide move to adopt K–12 curriculum reflective of the Standards.

There are two chapters categorized as *Social Inquiry*. Both of these authors use their classrooms as contexts to investigate larger social issues. Janelle McFeetors analyzed the construct of "success" and what it meant for her students—who had previously been unsuccessful in mathematics by standard measures of achievement—to be successful in mathematics class. She listened to what her students could tell her (and us) about the nature of their success in mathematics class. Jesse Solomon interviewed his students to learn more about the interactions of culture and pedagogy in mathematics class. The results of their work provide commentaries on our society and raise questions that help us think about external factors that affect classroom teaching.

Four chapters were *Part of a Larger Project*. The chapters by Nicole Garcia and Patricio Herbst, and Apolinário Barros and Dorina Sackman illustrate how involving teachers as participants in research projects can create opportunities for them to engage in their own research. Garcia was the participant in research orchestrated by Herbst, a university researcher. In their chapter we see how Herbst's research served as a context for Garcia to ask her own questions—about teaching mathematics with problems and issues of fairness—from the perspective of a teacher-researcher. In Barros's chapter with coauthor Dorina Sackman, we hear how the questions that Barros generated as a result of his participation in the larger research project affected the direction of that project. The chapters by Maureen Grant and Rebecca McGraw, and Michael Verkaik and Beth Ritsema provide examples of teachers and project directors becoming co-researchers. McGraw's invitation to Grant to participate in her dissertation research on classroom discourse led to a depth of collaboration that neither of them anticipated. Verkaik and Ritsema's chapter reflects a multi-level collaboration that resulted from a synergy among professional development, classroom change, and conversations about teaching in the context of implementing a reform curriculum.

The work reported on in these chapters represents teacher research as described by Brown, "Teacher research is a method of gaining insight from hindsight. It is a way of formalizing the questioning and reflecting we, as teachers, engage in every day in an attempt to improve student learning" (as cited in Dana & Yendol-Silva, 2003, p. 5). In the next section we turn to methods the authors used to carry out this work.

Methodology

As has been the case elsewhere, it is fair to say that the teachers in this book "built their research methodology upon the reflective practices they already engaged in as teachers" (Dadds & Hart, 2001, p. 145). Furthermore, the research methodology used in the work reported here is clearly rooted in classroom practice. Although Harrington and Huhn, Huhn, and Lamb interviewed students outside of class, they used mathematical problems as the basis for their work that were directly connected to the students' course experiences. Solomon's interviews were outside the realm of classroom instruction, but focused on investigating students' perceptions of their experiences in mathematics class and how those experiences related to their experiences in other out-of-class activities. Hendrickson and Christensen collected a range of data from assessments and interventions that they used with their own and their colleagues' students as they tested their hypotheses about parallels between student learning at the elementary and high school levels.

The remaining nine chapters are based on data that was collected in the normal course of instruction, though the unit of analysis and the number of data sources varies. Three chapters considered a semester-long course. Robinson looked at her teaching in an Algebra II course and used a departmental final exam as an external measure of improvement. Grant and McGraw used a first-year algebra course as a site for investigating their questions about classroom discourse. Their data analysis focused primarily on audio and videotapes of classroom instruction. McFeetors' work on success took place in what is often referred to as a general mathematics course. Using the research methodology of narrative inquiry (Clandinin & Connelly, 2000), McFeetors' data was the work and words of the students as she encouraged them to think about narratives of their success.

Three chapters collected data in the context of a unit of study. Zimmerman used a teaching experiment methodology (Cobb, 1999; Cobb, Confrey, deSessa, Lehrer, & Schauble, 2003) to investigate both the teaching and learning of a unit on probability simulation. Garcia focused her attention initially on an area unit that was part of Herbst's larger study. The student work, videotapes of classroom instruction, and student interviews collected as part of the study in which Garcia was a participant-researcher during her student teaching experience served as a backdrop for her to investigate the issues she found most relevant to her development as a teacher. Furthermore, she took a research stance into her future teaching as she continued to consider the issues in a new context. Glanfield, Baczuk and Oviatt used a unit on complex numbers and quadratic equations as the context for collecting student work and videotaping student presentations for their investigation into students' mathematical communications.

In three chapters, the work focused on one lesson. Carter, Gammelgaard and Pope report on their experience using a lesson study process to develop, implement, and revise a lesson on exponential functions. They also used existing teacher evaluation methods to assess ways in which participating in the lesson study process affected novice teachers' development. Barros and Sackman share their analysis of a lesson that began a series of activities on quadratic functions and parabolas. They use this lesson to illustrate the level of interaction among students and Barros's role as a facilitator of the classroom discussions—things he attributes to his participation in a research collaborative. Verkaik and Ritsema take a time-series approach to a lesson on equivalent rules and equations, videotaping the lesson twice over a span of two years. This allows for rich analysis of the challenges of implementing and sustaining instructional changes reflective of the *Standards*.

One thing that is striking about the work of the researchers in this book is the fact that they all valued what the students had to say and often sought answers to their research questions in the words of the students themselves. McFeetor does this explicitly when she talks about "listening" as a "methodological framing" and the importance of listening to the voices of students to draw meaning from their experiences. Solomon's study relied completely on his ability to listen to what his interviewees were willing to say to him. Huhn, Huhn, and Lamb took great care to minimize the stress of their interview situation so that they would have something to listen to.

Glanfield, Baczuk and Oviatt "listened" to students through their written and oral work in class. Many authors used the words the students spoke during videotaped class sessions as a way to gain insight into their mathematical thinking and to assess the effectiveness of the teaching. This emphasis on students and their words reflects a student-centered approach to research that is prevalent across the chapters in this volume.

Research Issues

There are several research issues illuminated by these chapters. For example, Zimmermann addressed issues of validity by making use of what she calls a "witness"—a knowledgeable person other than the researcher who served as "another set of eyes." McFeetors' chapter provides us with an example of theory building as she develops ideas about what she names "the emergence of voice." Both Harrington, and Hendrickson and Christensen extended the research of others in very deliberate, but different, ways. Harrington based his work on a classical literature review and replicated some of the ideas from earlier studies in an updated, technology-rich environment. Hendrickson and Christensen modified both the tasks and the grade level, but replicated the underlying mathematical structure of earlier work.

Although all of the authors apply the research of others to their situations to some extent, Robinson and Carter, Gammelgaard, and Pope's chapters provide examples where doing so forms the core of the authors' work. That is, they used what they had learned from reading research to inform their classroom and professional development practice. Solomon uses the work of others as a basis for examining assumptions underlying his teaching approach and providing a starting point for his own research.

There are several ways that engaging in research was connected to professional development. Verkaik and Ritsema provide an example of research as professional development. The process of analyzing videotapes of Verkaik's classroom practice was professional development, both for him and his colleagues. In his chapter with co-author Sackman, Barros discusses the way that being part of a formal research collaboration that met over an extended period of time provided him with a powerful form of ongoing professional development. In their chapter, Huhn, Huhn, and Lamb relate the professional conversations that they had with colleagues as a result of their research. Garcia and Herbst's chapter addresses research as self-reflection; Garcia makes the point that self-reflection is advocated by teacher education programs, but difficult for new teachers to begin. Based on her experience, she advocates engaging in research as a way for new teachers to develop productive reflective habits.

Although collaboration of some sort is a thread through all the chapters, two chapters specifically address the forming and negotiation of research collaborations. Grant and McGraw, a classroom teacher and university researcher, became teaching and research partners, both willing to have their teaching practices examined as part of the research process. They speak to the professional development power of working closely with a like-minded colleague on a particular aspect of practice. Glanfield, Baczuk, and Oviatt, a teacher educator and two classroom teachers, provide us with an inside look into how their research collaboration formed and the genesis of their joint work. This work allowed them to implement ideas that they had learned in other, less intensive forms of professional development—another reoccurring theme throughout the book.

WHAT WE CAN LEARN ABOUT RESEARCH FROM THIS BOOK

This book, and the series of which it is a part, allow us the opportunity to step back and reflect on what we can learn about research from a group of teachers who have engaged in research. This section reflects on five areas of insight: (a) the importance of collaboration and participation in communities that value research, (b) the potential of teacher research as a way to warrant teacher practice, (c) the power of video and other artifacts of teaching to support classroom inquiry, (d) connections between teaching and research, and (e) the publication process as additional professional development.

Importance of Collaboration and Community Participation

Although only a few of the chapters in this book focus explicitly on the development of a research collaboration (Grant & McGraw; Glanfield, Baczuk, & Oviatt) or participation in a research collaborative (Barros & Sackman), some sort of collaboration is a theme that runs through almost all of the chapters. In some cases, mention of participation in a research community is given as a reason for a teacher rethinking his or her ideas about practice (Verkaik & Ritsema; Garcia & Herbst); in others, it is a footnote of acknowledgement for the genesis of the current research (Solomon). As one reads the book, it is striking that the authors all had some connection to an active research community, often at a university. In other words, although the work stands alone, the authors were supported in some way by others engaged in systematic inquiry into classroom practice. This makes sense given the special skills involved in research and the need for researchers to be able to discuss their ideas with knowledgeable others.

Thie importance of being connected to other researchers speaks to the need for the development of structures—communities of inquiry—to support teachers as they engage in systematic inquiry in their classrooms. It is clear that these communities can take on many forms and still be successful. For some, it was a group of peers who engaged with university faculty through a graduate course (Hendrickson & Christensen; Robinson); for others it was a co-researcher (Grant & McGraw) or a group of colleagues (Verkaik & Ritsema; Huhn, Huhn, & Lamb). In some cases, a key component of the authors' work was the development of a group of teachers who were intentional about using research as a basis for working together in an extended way to improve the teaching in their school systems (Carter, Gammelgaard, & Pope; Henrickson & Christensen). For others, such a group provided the foundation for the work (Barros & Sackman; Garcia & Herbst). The need to collaborate was part of the research design in cases where a university researcher sought a teacher's classroom to use as the site for research (Garcia & Herbst; Grant & McGraw) and in Zimmermann's work through her "witness."

One of the ways that universities can support the budding teacher research movement is by offering graduate courses that focus on classroom research and provide teachers with both research skills and an important community of fellow inquirerers. Another is by valuing the work of faculty who serve as collaborators with or in support roles to teachers engaged in research. We see in the case of Grant & McGraw that there are benefits to merging the roles between classroom teacher and university researcher in these types of collaboratives. Many current tenure and promotion structures do not provide adequate credit for this kind of involvement and need to be modified if we are to maximize the benefits of this type of work for the field.

Likewise, school structures need to be changed to support teachers' engagement in inquiry as a natural part of practice. For the most part, teaching continues to be an isolating activity, with few opportunities to participate in meaningful discussions with other professionals. The move to establish teacher learning communities in school districts is certainly a step in the right direction. The next step is to ensure that these learning communities are grounded in existing research and positioned to engage in systematic inquiry into classroom and school-based practices.

There is one caution regarding collaborations and research communities; a theme that runs throughout the discussion of these relationships in the chapters in this book is the importance of respect. Respect for the people with whom one is collaborating and in community with was at the core of the relationships reported in this volume—particularly when those relationships crossed boundary lines, as is explicitly discussed in Grant and McGraw's chapter. Thus, it is not sufficient to create communities and form collaboratives; attention must be paid to the different perspectives and input each participant brings to the relationship and to honor those contributions accordingly.

Potential to Warrant Teaching Practices

The current "best practices" movement puts an emphasis on teaching actions that lead to student achievement that is easily measured by standardized tests. The research reported on in this book provides a richer picture of teaching practices and some insight into how they might be warranted. Historically, there has been a "preoccupation on the part of the teacher research movement with professional development and changed practice rather than public knowledge" (Ruthven, 2005). Pushing the work to be shared with the public, the goal of the Teachers Engaged in Research series may be a step towards providing warrants for teaching practices that produce student growth and wider-ranging achievement than that which has typically been the focus of public discourse. In this volume we see two distinct ways in which that could be done. Robinson provides a warrant for the changes in her practice by the improvement in her students' scores on an existing departmental final exam. McFeetors challenges existing measures and provides an alternative framework for assessing success. Clearly there is much potential for teachers to contribute to the public discourse on "best practices" and to provide warrants.

Power of Video and Other Artifacts of Teaching

The use of video in teacher education and professional development has recently been heralded in Brophy (2004). Perhaps not surprisingly given the interconnections between professional development and teacher research, eight of the thirteen chapters explicitly used videotapes to facilitate their research work. Verkaik and Ritsema captured the value of videotape in this quote from their chapter: "What a learning tool! Imagine being able to back up a conversation, slow it down, and think about what the other person was saying. ... I need not rely on a fleeting memory of the class period." The ease with which high quality videotapes can be made has opened up new avenues for systematic inquiry into classroom practice. Furthermore, providing dialogue from classroom episodes to illustrate the analysis and conclusions drawn from it holds promise as a means to increase transparency of the research process.

In other studies, student work and other artifacts of teaching play this same role. Sharing examples of student work or excerpts from a lesson plan to illustrate a point makes it much more likely that those reading the text will be able to engage in meaningful discussion rather than risk using the same words, but meaning different things. The groundedness of the chapters in specifics of practice—in evidence—increases their potential to make powerful contributions to mathematics education discourse.

Connections Between Teaching and Research

Throughout this book there is evidence of the symbiotic nature of research that teachers engage in—both regarding questions and methodology. The initial questions rose from the teachers' classroom experiences; both the findings and the process informed their classroom practice and raised additional questions. The work reported on here supports Marilyn Cohran-Smith and Susan Lytle's 1999 assertion that "inquiry both stems from and generates questions" (p. 21). Additional questions or "wonderings" were natural outgrowths of both the process and the findings for these researchers. In addition, the research process itself engaged teachers in behaviors that they were seeking from their students. Carter, Gammelgaard, and Pope put it this way in their chapter: "As we worked to design a lesson in which students were actively engaged, sharing ideas, listening carefully, making conjectures, and collecting data to inform decisions, we were doing exactly that ourselves—as a group of teachers we were engaged in our own professional learning." Furthermore, aspects of the research process—listening, observing, analyzing—are tied to teaching skills and improving a teacher's skills in either teaching or research seems likely to transfer to the other. This is particularly promising given that a frequent finding across the chapters was how difficult it was for teachers to move from a "telling" to a "listening" stance in their teaching.

There is evidence throughout this book that the research process oriented the teachers who engaged in it towards their students' thinking as the teachers sought to make sense of what was happening in their classrooms in deep and profound ways. An obvious way in which this occurred was through the methodological choices the teachers made. Over and over the choice of methodology was rooted in the teachers' classroom experiences; in turn, the focused application of these techniques during their data collection further developed skills used in their classroom instruction. It is the symbiotic nature of the work that seems to make teacher research different from that conducted by a university researcher who is not also a teacher. It also raises the question of whether university researchers are actually teacher-researchers whenever they engage with their research in such a way that it affects their own classroom teaching.

The Publication Process

A question worthy of investigation is what teachers learn through the editing process of publishing their research. It seems that the thinking and rethinking about one's work initiated by engaging in scholarly conversations with reviewers and editors may be an important form of professional development for teachers. The practical question is how to make this more viable. It became clear during the development of this book that the teachers who are engaged in research are often also the teachers who are leading their departments, initiating curricular change, and participating in many other professional activities that put demands on their time. Finding time to do the research is difficult enough; adding time to engage in post-research activities makes the process even more daunting. Another obstacle is the often slow and laborious publishing process. It is worth considering how technology can be used to accelerate the process and allow teachers' work to be published before the research seems like a distant memory.

CONCLUSION

The teachers whose work comprises this book found the time to do the research and endured the publishing process. By opening their analysis of their classroom practice to our inspection—by making their thinking transparent—these courageous teachers have invited us to think along with them and to learn more about our own teaching as a result.

By sharing their work, they have given the mathematics education community an important opportunity. It is my sincere hope that everyone who reads this book will be inspired—teachers, researchers, teacher-researchers, policy makers, administrators, and others interested in mathematics education. We can all learn from the findings and the light that they shed on issues important to mathematics education. We can also all learn from the kinds of questions the teachers asked and the process they went through to reach their findings. Often the teachers found that the answers they sought faded in importance as they became engaged in the process and learned things that they hadn't anticipated. May this also happen to you, as readers of their work.

NOTE

1. This chapter was written, in part, as a result of collaborative conversations with the editors of the other volumes of this series and the series editor. I wish to acknowledge the contributions of Cynthia Langrall, Stephanie Smith, Joanna Masingila, and Denise Mewborn.

REFERENCES

Clandinin, D. J., & Connelly, F. M. (2000). *Narrative inquiry: Experience and story in qualitative research*. San Francisco: Jossey-Bass.

Cohran-Smith, M. & Lytle, S. L. (1999). The teacher researcher movement: A decade later. Educational Researcher, *28*(7), 15-25.

Cobb, P. (1999). Individual and collective mathematical development: The case of statistical data analysis. *Mathematical Thinking and Learning, 1,* 5-43.

Cobb, P., Confrey, J., deSessa, A., Lehrer, R., & Schauble, L. (2003). Design experiments in educational research. *Educational Researcher, 32*(1), 9-13.

Dadds, M., & Hart, S. (2001). *Action research in education*. New York: Routledge/Falmer.

Dana, N.F., & Yendol-Silva, D. (2003). *The reflective educator's guide to classroom research: Learning to teach and teaching to learn through practitioner inquiry*. Thousand Oaks, CA: Corwin Press.

Lester, F. K. (Ed.) (in preparation). *Handbook of research on mathematics teaching and learning (2nd edition)*. Greenwich, CT: InfoAge/National Council of Teachers of Mathematics.

National Council of Teachers of Mathematics. (2000). *Principles and standards for school mathematics*. Reston, VA: Author.

Ruthven, K. (2005). Improving the development and warranting of good practice in teaching. *Cambridge Journal of Education, 35*(3), 407-426.

CHAPTER 2

PROBABILITY SIMULATION

What a Teaching Experiment Revealed About Student Reasoning and Beliefs

Gwendolyn Zimmermann
Hinsdale Central High School

Ten years ago, I was drawn into teaching from the corporate arena primarily for two reasons: (1) I wanted to make a difference in the education of students, and (2) I wanted a never-ending challenge. I realized that no matter how good or masterful a teacher I became, there would always be room for growth. In the first few years of my career as a teacher, I focused on my teaching skills, that is, the techniques I could learn to make my lessons more engaging. After about five years in the classroom, my attention shifted from "what was I going to do in the classroom" to "what my students would be learning." In other words, my focus changed from teaching to student learning.

At about this time, I applied to the doctoral program in mathematics education at Illinois State University. My course work helped to reaffirm my belief that in order to improve teaching, I needed to focus on student learning. I gained an appreciation for the supportive and guiding role

Teachers Engaged in Research
Inquiry Into Mathematics Classrooms, Grades 9–12, 19–38

research could play in my classroom. Through research, I could acquire knowledge vicariously and use this information to inform my mathematics instruction. I saw the value of helping classroom teachers appreciate and apply the insights research has to offer. However, I also noticed that many teachers feel disconnected from the research because they see little direct application to their classrooms. This seems to stem in part from the fact that full-time researchers who have conducted most of the studies are often far removed from the everyday experiences of the school setting. Thus, it became important for me to connect my experience as a classroom teacher and my new role as researcher. With this goal in mind, I was drawn to a teaching experiment methodology (Cobb, 1999) for my dissertation study. In the most simplistic terms, a teaching experiment is a cyclic process that uses previous research coupled with classroom-based analysis to drive instructional planning. This methodology allowed me to "experience, first-hand, students' mathematics learning and reasoning" (Steffe & Thompson, 2000, p. 267).

Having worked to develop an Advanced Placement (AP) Statistics course as part of my school's curriculum, I was interested in learning more about student thinking as it related to statistics. I decided to explore how students reasoned about probability simulation, a topic required by the AP Statistics curriculum and advocated for in the NCTM *Principles and Standards* (2000). Specifically, my research focused on two questions: (1) How do students reason about probability simulation in the context of an instructional unit? and (2) What student beliefs related to probability simulation emerge during instruction?

THE CONTEXT AND PARTICIPANTS

During the study I was a full-time teacher in a Midwest suburban high school with approximately 2,400 students. I was the teacher and researcher for one of two AP Statistics classes offered in the school. The prerequisite to enroll in the class was a grade of "C" or better in an Advanced Algebra course. Thus, the level of mathematical ability of the students varied from average to high-level ability. Three of the students were at the junior level and were taking AP Statistics concurrently with an honors-level precalculus course. The remaining 20 students were senior-level students having three years of mathematics prior to their senior year. Six of the 20 seniors were concurrently enrolled in an Advanced Placement Calculus course, and one senior was enrolled in honors precalculus. The student's previous courses determined the level of probability instruction he or she had prior to this class. Thus, the extent of probability instruction ranged from simple probability problems involving tree diagrams and basic counting principles to introductory instruction involving permutations and combinations to a unit involving conditional probabilities.

A critical component of the research study was what we called the *witness*. The role of the witness was to help me plan, to provide another set of eyes in the classroom for each of the 12 instructional days, and finally, during the analysis phase, to help validate my findings. In designing my study, I wanted a witness who knew the mathematics, who was familiar with the high school classroom, and who could be available every day for about two weeks to sit in the class and then meet after school to debrief and plan. My witness was a mathematics teacher who had just retired from the school district in which I taught. She both fulfilled all my requirements for a research witness and enjoyed the research process and her responsibilities.

THE TEACHING EXPERIMENT

The teaching experiment consisted of twelve 55-minute instructional sessions. Table 1 outlines the concepts covered over the course of the 12-day teaching experiment. Prior to each session, a hypothetical learning trajectory (Simon, 1995) was developed. The hypothetical learning trajectory consisted of learning goals, instructional activities, and conjectured learning processes. The learning goals were the targeted learner objectives for students, and the instructional activities were the problems and related activities used to assist students in meeting the learning goals. As the witness and I planned the hypothetical learning trajectory for each day, we hypothesized how students would approach problems, mistakes they might make in their reasoning, and hurdles that students may need to overcome to meet the desired learner outcomes. These hypotheses were the conjectured learning processes. Beginning with the learning goals, an instructional activity was developed for each day using knowledge gained from the previous day and research related to student reasoning about probability. These instructional activities and the main concepts they addressed are identified in Table 2.1.

To illustrate the hypothetical learning trajectory, I will use the first lesson as an example. I wanted to begin the unit by focusing on the five components of simulation listed in Figure 2.1. As a result, the following learning goals were identified for Session 1: (1) to construct a valid probability generator, (2) to determine and explain an event or trial for the simulation, (3) to use the probability generator to simulate a 1-dimensional probability situation, and (4) to use the simulation results to compare with "expected" results.

To accomplish these goals, we selected an activity titled "Counting Successes: A General Simulation Model" (Scheaffer, Gnanadesikan, Watkins,

Table 2.1.   Outline of Instructional Sessions for Whole-Class Teaching Experiment

Session	Instructional Activity	Concepts
1.	Counting Successes	1D, Design Simulation
2.	True/False History Test	MD, Design Simulation
3.	Free Throw Shooter and Blood Bank	MD, Design Simulation
4.	Randomly-Generated Outcomes	1D, Randomness, Technology
5.	Designing Simulations on the TI-83	MD, Conduct Simulation, Technology
6.	Tree Diagrams	MD, Theoretical Probability
7.	Venn Diagrams – Day 1	Theoretical Probability
8.	Venn Diagrams – Day 2	Conditional Probability
9.	A "False Positive" AIDS Test	Theoretical vs. Empirical
10.	What is Random Behavior?	Representativeness
11.	What's the Chance?	Independence of Events
12.	Are These Simulation Designs Valid?	Validity of Simulation Design

Note: 1D: One-Dimensional Sample Space refers to probability activities or stimulations that involve performing one random experiment.

MD: Multiple-Dimensional Sample Space refers to probability activities or simulations that involve the performing of multiple random experiments or the performing of one random experiment multiple times.

& Witmer 1996) for the first day. For this activity, students were given a situation where a system was proposed to eliminate the need for coins within our current monetary system. Students were given parameters to conduct a simulation to determine if the proposed system was "fair." We chose this activity because it addressed my learning goals for students in Session 1

Simulation Process

1. State the problem and list any assumptions.
2. Assign random digits to model problem outcomes.
3. Define a trial.
4. Repeat trial many times.
5. Determine empirical probability.

Figure 2.1. Simulation Process (adapted from Yates, Moore, & McCabe, 1999).

and because the only prerequisite knowledge was a basic understanding of proportions. Using published research (e.g., Fischbein & Gazit, 1984; Fischbein, Nello, & Marino, 1991; Fischbein & Schnarch, 1997; Garfield & Ahlgren, 1988; Piaget & Inhelder, 1975; Shaughnessy, 1992; Zimmermann & Jones, 2002), my witness and I created a hypothetical learning trajectory. We hypothesized that students would have little difficulty creating a probability generator or using a random number table to generate outcomes for the simulation. However, we hypothesized that some students might be inclined to calculate the problem theoretically rather than empirically. These hypotheses guided me primarily before and during instruction. I began by having my students work through the activity in groups. Because of the difficulties we hypothesized students might encounter, I carefully monitored student verbal and written responses to find evidence of students who might attempt to calculate the theoretical probability rather than do the simulation. Once the groups were finished, I facilitated a whole-class discussion posing questions to encourage students to share their strategies and reflect on the reasoning of other students.

At the end of this first day, the witness and I met to discuss our observations in the context of the hypothetical learning trajectory. In general, we found students had little difficulty with the activity; they were able to demonstrate sufficient knowledge of the concept of probability simulation and to use a random number table to conduct a simulation, as was anticipated. However, students did not attempt to use theoretical probability to address the problem posed in the activity. Based on these observations, we felt that the second day should extend student knowledge of probability simulation by presenting an activity involving a multi-dimensional situation. Planning for subsequent days followed a similar process.

DATA SOURCES AND ANALYSIS

A variety of data sources, both quantitative and qualitative, provided me with a rich picture of how student reasoning about probability simulation evolved over the 12-day instructional unit and revealed patterns helpful to instructional planning of probability simulation. Using both types of data sources provided a wider lens into student reasoning than either one alone could provide.

Students were given a pre-, post-, and retention assessment. The preassessment was administered two weeks before the instructional unit began. The postassessment was given immediately following the conclusion of the teaching experiment, and students took the retention assessment four weeks after the completion of the instructional unit. The three assessments contained parallel items designed to assess students' abilities on two different levels. One level was to determine a student's ability to assess the validity of a given probability simulation, and the second level was designed to assess a student's ability to construct a valid probability simulation. Using these assessments, two kinds of quantitative analysis were completed: one that compares the total scores on each of the three tests and another that examines the frequency of responses related to each component of a simulation. Figure 2.2 contains the first part of the preassessment, "The Pizza Problem," as an example of the types of questions students were asked on each assessment instrument.

The qualitative data that were gathered came from three sources. The first source was the written responses to the assessment instruments. The second source was student interviews. Four students were purposefully sampled (Miles & Huberman, 1994) to provide a more in-depth look into individual student thinking about probability simulation. More explicitly, data from the preassessment were used to help identify students with varying abilities in solving probability simulation problems. I also intentionally chose students I felt would best be able to articulate their thinking when interviewed. These students were interviewed after the preassessment and again after the postassessment. The third source was audio and videotapes of each of the 12 instructional sessions. Two videotapes and six audiotapes were recorded for each instructional session to view the classroom from two different perspectives and to preserve the discussions of each student group.

To help make sense of the qualitative data I collected, I used Miles and Huberman's (1994) "three part analysis" (pp. 10–11) to carry out a qualitative analysis of the whole-class teaching experiment. The first part was *data reduction*. During this stage all the data sources were coded to generate images and impressions of individual and collective student reasoning and

TASK 1: PIZZA PROBLEM

A student, Stan, was given the following problem.

The Pizza Wagon has determined that 60 percent of their phone orders for pizza contain meat (sausage, pepperoni, etc.) and the remaining 40 percent of their phone orders are for pizzas with no meat (cheese, veggie, etc.). What is the probability that the next two phone orders for pizza are each with meat?

To simulate the Pizza Wagon's situation, Stan used colored chips. Stan chose 6 red chips to each represent an order for pizza with meat, and he chose 4 green chips to each represent an order for pizza without meat. To simulate the actual order, Stan put all 10 chips into a bag, shook the bag, and drew out one chip. He recorded the color, put the chip back, and then repeated this action a total of 50 times.

1) Remembering that the Pizza Wagon is trying to determine the probability that the next 2 pizzas have meat, <u>do you think that Stan's simulation would enable him to determine the probability that the next two pizzas have meat</u>? Justify your response.

2) If you don't think Stan's simulation will work, how would you change the simulation to determine the probability that the next two pizzas have meat?

3) Suppose Stan conducted his experiment 50 times and his results were as follows:

RRRGRRRGGGGRGRRGGGGRRRGRGRRGGGRRGGGRRGGR-RRGGRGRGGR

Red chip	25 times	
Green chip	25 times	

Using the outcomes or the results from Stan's experiment, could you determine the probability that the next two phone orders for pizza have meat? If your response is "yes," calculate the required probability and explain your reasoning. If your response is "no," explain why not.

Figure 2.2.   Example of Assessment Item

beliefs. Audio and videotapes were analyzed by viewing the daily activities of each group in chronological order (Cobb & Whitenack, 1996). Key episodes were then used to make initial conjectures about student reasoning and beliefs related to probability simulation that were modified based on analysis of subsequent classroom teaching episodes.

The required components in the simulation process (see Figure 2.1) provide the frame I used for analyzing students' responses to simulation tasks. In particular, as I analyzed the audio and videotapes, I coded dialog according to the simulation component. If a student was discussing how he or she would assign numbers to simulate a problem, I would code this part of the dialog as related to the probability generator, the second step in the simulation process (see Figure 1). When students discussed how many times they ran a simulation, this would be coded as "repeated trials." This helped me to organize that data for further analysis.

The identification and analysis of students' beliefs about probability simulation was guided by Schoenfeld's (1985) definition of beliefs as a lens through which an individual sees and approaches mathematics, and by other research related to the topic (Fischbein & Gazit, 1984; Fischbein, Nello, & Marino, 1991; Fischbein & Schnarch, 1997; Garfield & Ahlgren, 1988; Piaget & Inhelder, 1975; Shaughnessy, 1992; Zimmermann & Jones, 2002). Specifically, beliefs were analyzed using codes such as the following: students said they "believed," students responded to a question that asked what they believed, or students added an afterthought to their response that reflected a disposition. Once a student's belief was identified, I then connected the belief to the relevant simulation component. In this way, I was able to begin to organize emerging beliefs.

The data reduction process described above was repeated for each of the six student groups. In the second part of the three-part analysis process, *data tables* were developed using the codes generated in the first part of the process in order to compare students' reasoning on each of the simulation steps across all qualitative data sources. The final part of the process involved *drawing conclusions and verifying results.* I was able to verify results through a negotiating process (Cobb & Bauersfeld, 1995) involving the witness. More specifically, I provided the witness with my initial findings, which she then validated using her field notes and video and audiotapes as needed. Any discrepancies between us were negotiated until a consensus was reached. After the coding process, the witness reviewed the data displays to determine if she concurred with the trends and patterns that I had identified. Any disagreement was discussed and the analysis was modified to our mutual satisfaction.

Students' Reasoning About Probability Simulation

One set of insights into student reasoning about probability simulation comes from their performance on the pre-, post-, and retention assessment instruments. The means and standard deviations of each of these assessments are presented in Table 2.2.

A multivariate test using Wilks's Lambda revealed significant differences between the mean scores on the three assessments ($p < .001$). Furthermore, the assessments were compared in pairs to examine changes between the three assessment points. These pairwise comparisons indicate that the postassessment scores were significantly higher than the preassessment scores ($p < .001$). Likewise, the retention assessment scores were also significantly higher than the preassessment scores ($p < .001$). There was not a significant difference between the post- and retention assessments ($p = .116$).

Comparing the postassessment with the preassessment scores revealed that the mean score increased by over 75% while the standard deviation decreased by 3.6 points (see Table 2.2). These figures represent a significant increase in student performance from the preassessment to the postassessment, and they also indicate a more consistent performance at the time of the postassessment. What these data indicate is that students' reasoning about probability simulation increased significantly after the whole-class teaching experiment and students maintained the higher level of reasoning six weeks after the teaching experiment when the retention assessment was administered.

Probability simulation can be separated into discrete processes as identified earlier in Figure 2.1. Using an adaptation of the simulation process as defined by Yates, Moore, and McCabe (1999), I examined student responses to the three assessment instruments for frequency of valid responses on each of the major components that constitute a probability simulation. As illustrated in Table 3, in all but one of the components (assumptions in the construction task), more students demonstrated the

Table 2.2.   Means and Standard Deviations for the Pre-, Post-, and Retention Assessment Scores.

Assessment	M	SD	N
Pre-	10.2 (49%)	5.7	21
Post-	17.9 (85%)	2.9	21
Retention	18.9 (90%)	2.1	21

ability to reason about probability simulation after instruction. To be more precise, significant progress was made by students in both their ability to use simulated outcomes to determine the probability of a situation and to recognize the effect of repeated trials on the empirical probability.

In general, the qualitative data both confirmed the findings of the quantitative data and provided more detail into student reasoning. Following is a summary of some of the findings from the analysis of the qualitative data sources organized by simulation component (see Figure 2.1).

Analysis of the preassessment indicated that although students made *assumptions* about the simulation problems (see Table 2.3), these assumptions were idiosyncratic in nature. However, after the instructional unit, even though students made fewer assumptions, the assumptions students referred to were more normative. (By normative, I mean relevant to the design of the simulation, such as assumptions about the independence or dependence of events.) For example, consider Lacey's idiosyncratic response to the pizza problem on the preassessment (see Figure 2.2): "I would consider the day and season it is and if any special orders are made

Table 2.3.   Valid Student Reasoning in Components
of a Simulation by Assessment

	Number of Students (n = 21)		
Process in a Simulation Problem	*Pre-*	*Post-*	*Retention*
Evaluated a Probability Simulation			
Assumptions	1	5	3
Evaluated probability generator	18	21	21
Recognized need for 2-D trial	14	21	21
Accepted randomness of outcomes	7	18	18
Calculated empirical probability given the outcomes	7	18	18
Constructed a Probability Simulation			
Assumptions	5	1	2
Constructed probability generator	12	19	19
Constructed 2-D Trial	11	21	19
Calculated empirical probability	13	19	21
Repetition of trial	6	16	14

that day." Following the teaching experiment, the normative assumptions related mainly to maintaining a valid probability generator by replacing a drawn chip and to the independence of events. Analysis of the videotaped classroom sessions revealed that I spent little instructional time discussing the role of assumptions in the simulation process; this likely contributed to the lack of student attention to assumptions when constructing a probability simulation.

Students became more versatile in their ability to recognize and *construct a valid probability generator.* Research findings indicate that the construction of a one-dimensional probability generator is accessible to most students (Benson, 2000; Benson & Jones, 1999). The findings of my study demonstrate that, although students were generally able to recognize a valid probability generator, the construction of a valid two-dimensional probability generator proved to be more difficult, confirming earlier research findings (Zimmermann & Jones, 2002). For example, one problem on the preassessment provided probabilities of the DJ of a particular radio station playing hip-hop, alternative, and country music and inquired about the probability of hearing a hip-hop song both times when turning on the radio at 10 am and 2:30 pm. When asked to construct a probability generator for this problem, three students responded that they would simply turn on the radio. Breanna was one such student. She wrote, "My simulation would be to turn on the radio at 10:00 a.m. and then again at 2:30 p.m. for every day for 4 weeks straight. At each day, record the type of music played at 10:00 a.m. and 2:30 p.m. as a set."

Following the teaching experiment, student reasoning evolved to a point where students were able to construct valid probability generators for multidimensional trials and trials involving dependent events. For an example of a valid probability generator, consider Cade's approach to the space shuttle problem. He used a combination of a calculator and a spinner to construct a probability to determine the probability that both engines of a space shuttle would fail given the probability engine one would fail was 0.2 and the probability engine two would fail was 0.3.. He said he would "take a calculator and have it choose a random integer from 1-10: (for S1) 1-2 was failure, 3-10 [engine] works. If you get a 1-2, use a spinner divided in 10 spaces, 1-3 failure, 4-10 works."

In Session 1 students were taught to use a random number table. During Session 2 students were introduced to the random integer capability of the graphing calculator. From that day forward, the calculator became the tool students used to generate random numbers in class. At one point I tried to integrate the use of spinners, but the students seemed to prefer using the calculator. However, they showed versatility in using other devices, such as spinners, balls, and chips, in the post- and retention assessment problems. It is unclear why students demonstrated such versatility in their ability to use a variety of probability generating devices. However, I conjecture that the use of the calculator to conduct simulations played an important role. The calculator syntax required that students focus on two specific components of a simulation, namely the assignment of random digits and the trial. In using the graphing calculator to conduct probability simulations, students likely developed a deeper sense of what was needed to produce valid probability generators regardless of the device.

When students' ability to *define a valid simulation trial* was analyzed, invalid reasoning was identified on each of the three assessments. Students' difficulty in being able to define a valid trial resulted from their inability to use proportions correctly and to recognize a valid sample space, especially in the case of two-dimensional trials. More specifically, students transformed a two-dimensional trial into a one-dimensional trial. For example, on the space shuttle problem mentioned earlier, Kacy explained that she would use "randInt [random number generator] on my calculator and use the variables 1 for S1 and 2 and 3 for S2. If I got a 1 that would mean S1 failed, and if I got a 2 or a 3 that would mean S2 failed." Kacy seemed unaware that her design did not represent the sample space in the problem. In contrast, Cade, as mentioned earlier, demonstrated valid reasoning when he recognized the multi-dimensional nature of the space shuttle problem and accurately accounted for two engines.

The fourth component of the simulation process requires that students *repeat a trial many times.* When I examined student reasoning with respect to *repetition of trials,* students showed increased awareness that as the number of trials increased, the empirical probability approached the theoretical value. Thor was articulate in his explanation of the effect increasing the number of trials had on the empirical probability: "If you only do the experiment 50 times, you may not get a proper ratio that is accurate because there were only 50 trials. Each trial has less of an influence as you do more tests. If you do 100,000 tests, each test will have little change in the overall outcome but all the tests together give you a much more accurate representation of your true value." Although other students were not as articulate, many reasoned "more trials provided a more accurate empirical probability." While students did not always want to simulate a substantial number of trials, they were aware that the empirical probability showed less fluctuation and more stability as the number of trials grew.

After instruction all but one student was able to calculate an *empirical probability.* Students' ability to calculate the probability appeared to be connected to their ability to define a valid two-dimensional trial and then use their definition to calculate the empirical probability. As an example of student reasoning, once again consider the pizza problem on the pre-assessment. Students who reasoned validly on the post- and retention assessments used a strategy similar to the strategy Lacey used on the pizza problem. Lacey explained, "I broke up the count into two sections because the way I'd interpret the simulation and data would be to see if two chips were consecutively pulled out as meat pizzas. This happened 5 times out of the 25 'pulls.'" In other words, students broke the string of outcomes into pairs representing the two-dimensional trial. They then tallied the number of target pairs out of the total number of pairs to produce an empirical probability.

Some students in the preassessment exhibited an inability to determine the *empirical probability* because they seemed unwilling to accept the simulated outcomes as valid. Batanero and Serrano (1999) reported that high school students do not necessarily understand the nature of randomness. Results of my teaching experiment confirm that, although many students (over 85%) accepted the randomness of simulated outcomes, some students struggled with the concept. For those students who had difficulty with randomness, the representative heuristic (Kahneman & Tversky, 1972) was a barrier to accepting "unexpected" results. These students believed that the sample of simulated outcomes should match or be representative of the population probabilities of the original problem. Referring to the pizza problem on the preassessment, students who struggled with randomness did not accept the equal distribution of outcomes in the pizza problem because they believed the outcomes did not accurately reflect the problem situation. Mai's response was typical of these students: "In actuality their percentages of orders were for 60:40, not 50:50 like the outcome." This highlights a connection between reasoning and beliefs. Not only did my teaching experiment provide insight into student reasoning on probability simulation, it also revealed a number of student beliefs related to the concept of probability simulation, some helpful and some more problematic.

STUDENTS' BELIEFS ABOUT PROBABILITY SIMULATION

Students' beliefs can influence how they reason about probability simulation. That is, reasoning and beliefs together determine how a student approaches and thinks about problems. Therefore, one of my goals was to examine the beliefs of students that emerged as they reasoned about probability simulation. These beliefs were categorized as either helpful or problematic, as illustrated in Table 2.4.

Helpful Beliefs

Helpful beliefs are considered to be potentially beneficial to the learning of probability simulation. The helpful beliefs identified were generally directly related to simulation components, such as assumptions are necessary in carrying out a simulation, and the probability generator should correspond to the probabilities in the problem. Students seemed to recognize that assumptions were an inherent part of the process. However, as indicated earlier, students seldom made assumptions explicit, and those assumptions that were explicitly stated by students were often idiosyncratic. By the postassessment, such assumptions had become more normative, often referring to the independence of trials. As an example, when asked to assess a probability simulation design on the retention assessment, 12 of the 15 students responded that when a chip was drawn it should be replaced before drawing another chip. The remaining 3 students explicitly

Table 2.4.   Student Beliefs by Assessment

Belief	Number of Students Expressing Belief		
	Pre-	Post-	Retention
Helpful Beliefs			
Inherent assumptions in simulation model	8	8	15
Probability generator should correspond to given probabilities	18	21	21
Number of trials (n) that should be simulated			
• $n \leq 10$	6	1	0
• $0 < n < 100$	8	3	8
• $n \geq 100$	4	15	10
• Other	3	2	3
Context influences number of trials	0	5	0
Increased trials make empirical probability more "accurate"	7	16	15
Problematic Beliefs			
Theoretical probabilities can always be determined	2	0	0
Representativeness	10	3	0
Outcome approach	2	0	0

stated that the events were assumed to be independent. Other helpful beliefs were related to simulation trials. For instance, students believed that, although many trials provided a more precise empirical probability, there exists a difference between an ideal number of trials and a practical number of trials. That is to say, students seemed to agree that in theory many trials (in excess of 100) should be done, yet they believed 30 trials was enough for classroom purposes.

An unintended finding was that some students (5 out of 21) understood that the context of the problem might influence the number of trials and specifically referred to the context when justifying their answer. This finding only appeared in the postassessment, where students were asked how many trials should be conducted to simulate engines failing on a space shuttle. As one student responded, "As many as possible. This is a space shuttle. Lives depend on this value. You want it to be accurate to the nth degree." This seems to indicate how the context of a problem can influence student reasoning. Compared to the other assessment questions related to such things as music, pizza, and tennis, students seemed to strongly believe that more trials should be conducted for the space shuttle simulation, since this situation was of a very serious nature—involving human lives.

Problematic Beliefs

Three problematic beliefs were discerned. Beliefs are considered problematic if they may work to prevent a student from learning certain aspects of simulation because misconceptions are incorporated into that belief. Two of the problematic beliefs were reasoning heuristics that have been identified in other research: the representative heuristic and the outcome approach (Kahneman & Tversky, 1972; Konold, Pollatsek, Well, Lohmeier, & Lipson, 1993). Recall that on the preassessment the representative heuristic occurred when some students believed that the outcomes for the pizza problem were invalid because they did not represent the probabilities given in the problem. People who reason using the outcome approach believe they are asked to predict the next outcome rather than determine the probability of an event. Kacy's written response on the preassessment is indicative of this type of reasoning: "I predict a 30% chance of hearing another hip-hop song." The representative heuristic occurred on all three assessments but with decreasing frequency. The outcome approach was never identified after the preassessment. Neither of these beliefs occurred during instruction. One limitation of my research design was that it did not allow me to trace individual reasoning and beliefs. However, I conjecture that the classroom experiences provided students with opportunities to compare and discuss reasoning and thus helped students to correct earlier identified misconceptions.

The third problematic belief identified was the belief that the theoretical probability can always be determined. This belief was quite unexpected for two reasons. One reason was that I did not find any research to suggest students would exhibit such a belief. Secondly, as a teacher I thought the experience of constructing and doing numerous probability simulations would have helped students to develop a sense of purpose and value of the role of simulations. I was wrong. During one instructional session, when I asked what was the purpose of doing simulation, a student responded, "To see the difference between simulated and theoretical." As a follow-up question, I asked, "Can we always determine the theoretical probability?" Most of the students in the class held the belief that a theoretical probability could always be determined. This belief proved to be extremely resistant to instructional intervention.

SUMMARY

It was my intent to examine how students' reasoning and beliefs about probability simulation emerged during an instructional unit. My findings provide an in-depth view into students' reasoning. Across all the simulation components, students made significant progress. Assumptions became more normative. Students became adept at designing probability generators, and then using the probability generators to define two-dimensional trials. Over time, students overcame any bias against randomness and were able to use outcomes to calculate empirical probabilities. Students also developed an understanding of repeated simulation trials. They would often comment that one should continue to repeat trials "until it [empirical probability] stops fluctuating." Finally, my study revealed some of the helpful and problematic beliefs that emerged during an instructional unit on probability simulation.

IMPLICATIONS FOR CURRICULUM AND INSTRUCTION

The experience of designing, implementing, and analyzing an in-depth research project opened my eyes to all that I can learn and share with others. The results of my research demonstrate that probability simulation is accessible to high school students and that instruction can play a role in helping to enrich student understanding of simulation and related concepts, such as randomness, empirical probability, and theoretical probability. The process of providing students with interesting, contextual problems to simulate also encourages students to share and debate crucial concepts, which works to promote mathematical discourse and understanding. Furthermore, given the importance probability simulation plays in connecting empirical probability to theoretical probability, my findings suggest that it might be possible to include simulation earlier in the curriculum and not only within an AP course. Toward the end of the unit, my students were given simulation problems involving multidimensional trials and situations where the events were not independent. The ease with which students were able to simulate such challenging problems suggests that even less advanced students could benefit from studying probability simulation; in addition, AP students could be given more complex problems that require greater insight when defining a trial and decision-making related to concepts such as independence, dependence, and conditional probability.

On a more personal level, I learned many things that I intend to apply the next time I teach the AP Statistics course. In my study, students developed a more normative view of assumptions by the end of the teaching experiment. Although this was encouraging, it also revealed the narrow view students held about assumptions. I learned from my research experience that instruction should include a greater focus on assumptions, both explicit and implicit. Assumptions were not discussed in depth, and, as a result, students did not develop a conceptual understanding of their importance. If we had spent more time discussing the implication and importance of assumptions in a simulation context, they may have developed a better understanding of their role in the process.

I also learned that instruction should incorporate activities or problems that will help students develop an understanding of the variability in a simulation of only 30 or 40 trials. My students recognized that many trials were desirable, yet they seemed unwilling to actually simulate many trials, even with the speed of a graphing calculator. Thirty or 40 trials are not enough for students to recognize the inherent variability in the simulation process. Next time I will carefully choose the problems or applications so that students are more likely to recognize the importance of many trials. I may also ask students to explore the standard deviation as the number of trials increases. Finally, I would also like to explore ways to help students develop a greater appreciation for the value and usefulness of probability simulations by providing them with the opportunities to design and conduct simulations for real world problems.

This research project gave me an opportunity to focus extensively on one unit of study to see how my students reasoned and how my instruction affected that reasoning. By designing and implementing a teaching experiment to explore student reasoning, I was able to combine two of my passions: helping students learn mathematics and learning to become a better teacher.

ACKNOWLEDGMENTS

This research project was part of a doctoral dissertation completed July 2002 at Illinois State University. My committee consisted of chairpersons Graham A. Jones and Cynthia Langrall, and members Kenneth Berk and Edward S. Mooney. Funds from Illinois State University's Ora Bretall Scholarship Fund were used to help support this research project.

REFERENCES

Batanero, C., & Serrano, L. (1999). The meaning of randomness for secondary school mathematics. *Journal for Research in Mathematics Education, 30,* 558-567.

Benson, C. (2000). *Assessing students' thinking in modeling probability contexts.* Unpublished doctoral dissertation, Illinois State University.

Benson, C., & Jones, G. (1999). Assessing students' thinking in modeling probability contexts. *The Mathematics Educator, 4*(2), 1-21.

Cobb, P. (1999). Individual and collective mathematical development: The case of statistical data analysis. *Mathematical Thinking and Learning, 1,* 5-43.

Cobb, P., & Bauersfield, H. (Eds.). (1995). *The emergence of mathematical meaning: Interaction in classroom cultures.* Hillsdale, NJ: Lawrence Erlbaum Associates.

Cobb, P., & Whitenack, J. (1996). A method for conducting longitudinal analyses of classroom videorecordings and transcripts. *Educational Studies in Mathematics, 30,* 213-228.

Fischbein, E. & Gazit, A. (1984). Does the teaching of probability improve probabilistic intuitions? *Educational Studies in Mathematics, 15,* 1-24.

Fischbein, E., Nello, M. S., & Marino, M. S. (1991). Factors affecting probabilistic judgments in children and adolescents. *Educational Studies in Mathematics, 22,* 523-549.

Fischbein, E., & Schnarch, D. (1997). The evolution with age of probabilistic, intuitively based misconceptions. *Journal for Research in Mathematics Education, 28,* 96-105.

Garfield, J., & Ahlgren, A. (1988). Difficulties in learning basic concepts in probability and statistics: Implications for research. *Journal for Research in Mathematics Education, 19,* 44-63.

Kahneman, D., & Tversky, A. (1972). Subjective probability: A judgment of representativeness. *Cognitive Psychology, 3,* 430-454.

Konold, C., Pollatsek, A., Well, A., Lohmeier, J., & Lipson, A. (1993). Inconsistencies in studentsí reasoning about probability. *Journal for Research in Mathematics Education, 24*(5), 392-414.

Miles, M., & Huberman, A. M. (1994). *Qualitative data analysis.* Thousand Oaks, CA: Sage.

National Council of Teachers of Mathematics. (2000). *Principles and standards for school mathematics.* Reston, VA: Author.

Piaget, J., & Inhelder, B. (1975). *The origin of the idea of chance in children* (L. Leake, Jr., P. Burrell, & H. Fischbein, Trans.). New York: W. W. Norton, 1975 (Original work published 1951).

Scheaffer, R., Gnanadesikan, M., Watkins, A., & Witmer, J. (1996). *Activity-based statistics.* New York: Springer-Verlag.

Schoenfeld, A. (1985). *Mathematical problem solving.* Orlando, FL: Academic Press.

Shaughnessy, J. M. (1992). Research in probability and statistics: Reflections and directions. In D. Grouws (Ed.), *Handbook of research on mathematics teaching and learning* (pp. 465-494). New York: Macmillan.

Simon, M. (1995). Reconstructing mathematics pedagogy from a constructivist perspective. *Journal for Research in Mathematics Education, 26,* 114-145.

Steffe, L., & Thompson, P. (2000). Teaching experiment methodology: Underlying principls and essential elements In A. E. Kelly & R. A. Lesh (Eds.), *Handbook of research design in mathematics and science education* (pp. 267-306). Mahwah, NJ: Lawrence Erlbaum Associates.

Yates, D., Moore, D., & McCabe, G. (1999). *The practice of statistics.* New York: W. H. Freeman.

Zimmermann, G., & Jones, G. (2002). Probability simulation: What meaning does it have for high school students? *Canadian Journal Science, Mathematics, and Technology Education, 2,* 221-236.

CHAPTER 3

STUDENT UNDERSTANDING OF THE CONCEPT OF LIMIT IN A TECHNOLOGICAL ENVIRONMENT

William J. Harrington

State College Area School District
and the Pennsylvania State University

Teachers do informal research in their classrooms all the time. We try a new lesson activity, form of evaluation, seating arrangement, grouping of students, or style of teaching. We assess, reflect, modify, and try again as we consider the perceived consequences of the changes we made. Sometimes we share what we learned with colleagues in our school, or perhaps we might present a talk at a professional meeting. I believe that we can strengthen our profession by learning to be more deliberate about the informal research that we conduct within the walls of our individual schools and by being more purposeful about sharing what we have learned. To that end, I share this account of a research project related to the use of technology in calculus that I completed in my high school as part of a graduate class.

Teachers Engaged in Research
Inquiry Into Mathematics Classrooms, Grades 9–12, pages 39–58
Copyright © 2006 by Information Age Publishing
All rights of reproduction in any form reserved.

Computer and graphing calculator technologies have become commonplace in schools. The high school where I have taught since 1986 now issues graphing calculators like textbooks to nearly every student in the school. Calculator overhead projection is available in every mathematics classroom and several computer labs are available for class use. The regular availability of this sort of technology to students has forced a rethinking of the teaching and learning of mathematics. Instead of being regarded as mere supplementary tools, calculators and computers can be woven into the framework of the curriculum and instruction itself. The ways in which students use these technologies can have an impact on their mathematical thinking and understandings. The use of graphing technology allows for more conceptual approaches to learning, particularly a shift from obtaining graphs to interpreting graphs. The relationship between symbolic and graphical representations becomes much more accessible and students inhibited by the formal mathematics required for some problems can access solutions using graphing technology (Lauten, Graham, & Ferrini-Mundy, 1994).

The foundational calculus concept of limit has been a great challenge to many beginning calculus students. Because the concept of limit is generally encountered early in the study of calculus, students can be turned off to the subject as a whole if they find this concept to be too abstract to comprehend. Significant reform efforts in the teaching of calculus have been underway for some time now. Some in calculus reform believe that a formal definition of limit is inappropriate for beginning calculus students (Gass, 1992). Teachers of calculus must find ways to help students understand the concept of limit, and graphing technologies may be able to provide these beginning calculus students with the conceptual understanding they need and access to problem solutions that might otherwise be beyond their reach. Graphing calculators and computer algebra systems (CAS) can provide a means for an informal introduction and exploration of limit that builds on students' intuitions and can lead to a more formal definition of limit (Gass, 1992).

If graphing technologies are going to have a positive impact on mathematics pedagogy, however, careful evaluation is necessary. As curriculum and teaching changes that incorporate greater use of technology are implemented, we need to examine the effect on students' learning. How students develop meanings from activities within a technological setting must be considered. In particular, what understandings of limit do students develop from the activities and curriculum of a typical classroom where some form of graphing technology or CAS is used as a primary tool for learning the concept? Because my research did not take place within the classroom as concepts were being learned, I was not able to answer the question of *how* students' personal meanings of limit have been developed beyond speculation. Instead, I sought to examine *what* meanings of limit students developed.

WHAT WAS LEARNED FROM THE LITERATURE

Significant research has sought to examine students' concept images of limit (Davis & Vinner, 1986; Lauten, Graham, & Ferrini-Mundy, 1994; Szydlik, 2000; Tall & Vinner, 1981; Williams, 1989, 1991). One important claim is that "every concept has a concept image, namely, the way it is viewed by particular individuals" (Davis & Vinner, 1986, p. 285). The concept image consists of the cognitive structure in the individual's mind that is associated with a given concept. While examining students' understanding of functions and limits, Lauten et al. (1994) found that "students experience conflict between formal, precise definitions and the informal, natural language interpretations used conveniently in discourse" (p. 227). Mathematicians use phrases like tends to, approaches, converges, and limit interchangeably while each may have somewhat varying meanings to students from other contexts. Realizing that the concept image that students develop may not be the image that a teacher intends, I wondered what images of limit students taught using technology might hold.

In addition to the mathematically accurate formal or static notion of limit, research literature identifies several types of mathematically incomplete and inaccurate concept images held by students. These include limit as dynamic or motion, as unreachable, as an approximation, and as a boundary (Craighead & Fleck, 1997; Davis & Vinner, 1986; Gass, 1992; Lauten et al., 1994; Monaghan, 1991; Szydlik, 2000; Tall & Vinner, 1981; Williams, 1989, 1991). The various notions of limit identified in the research literature are characterized in Table 1. Williams (1989, 1991) studied the limit concepts of college calculus students and examined factors that inhibited students from adopting a formal static view of limit. He found that students' notions of limit often included two or more of the conceptions described in Table 1 and that they accepted different descriptions of limit as valid based upon these conceptions.

Having found such extensive research identifying these conceptions and showing that students seem to often hold multiple informal conceptions of limit, I wondered what was known about how informal teaching approaches to the concept of limit using technology might inhibit or encourage these various conceptions. Lauten et al. (1994) studied how students' understanding of function and limit had been influenced by the availability of a graphing calculator. Similar to findings in the Williams' (1989, 1991) studies, students' concept image seemed to shift with the con-

Table 3.1.   Limit Conceptions Identified in the Research and
Corresponding True/False Interview Questions

Dynamic-Theoretical	Boundary	Formal, Static	Unreachable	Approximation	Dynamic-Practical
Limit is dynamic, involving a notion of movement; describing how a function moves as *x* moves toward a certain point.	Limit is a boundary (asymptote) which can not be crossed.	Limit is a number that the y-values of a function can be made arbitrarily close to by restricting *x*-values	Limit is "as *x* approaches *s*" so it never really gets there.	We can only approximate *s* "as *x* approaches *s*" since the function never really gets there.	Limit is dynamic, involving a notion of movement; plugging in numbers closer and closer to a given number until the limit is reached.
			True/False?		
			A limit is a number or point the function gets close to but never reaches.	True/False?	
True/False?	True/False?	True/False?		A limit is an approximation that can be made as accurate as you wish.	True/False?
A limit describes how a function moves as *x* moves toward a certain point.	A limit is a number or point past which a function can-not go.	A limit is a number that the y-values of a function can be made arbitrarily close to by restricting *x*-values.	A limit is a number that the *y*-values of a function can be made arbitrarily close to by restricting x-values.		A limit is the value reached by plugging in numbers closer and closer to a given number.

text of the task presented. These researchers found only minimal unprompted use of available graphing calculators to solve problems, though this could be because subjects did not have significant experience with this technology—some as little as two weeks. They believed that the tracing operations that subjects performed in seeking limits of given functions seemed to be associated with a dynamic view of limit, as subjects described limits with motion references.

The need for more research in this area was clear to me. Lauten et al. (1994) made calculators available to students, but these students were unfamiliar with the technology and did not use it significantly. I wondered what I might learn if I examined the conceptions of our students who were all quite familiar with graphing calculator technology. Because I wanted to investigate the ways in which the students I interviewed would make use of technology, I provided them with limit problems that exceeded their analytical skills. I wanted to see how the students would use the technology, whether they could find limits successfully with the technology, and what concept images they held with respect to limit. I thought that a determination of a student's typical use of technology and their personal understanding of limit might reveal links between the students' preferred methods of solution and their particular understanding of the limit concept. I also thought that an examination of the accuracy of students' solutions along with their personal concept images would provide evidence of whether students' ability to determine limits using informal techniques with technology is good evidence that they understand the concept. I focused my work on two main questions: (1) What is the relationship between students' solution methods and their conceptions of limit? and (2) What is the relationship between students' ability to successfully use technology to determine limits and their conceptual understanding of limit? I believed that information with respect to these issues and answers to these questions could be important for calculus teachers to know.

RESEARCH PROCEDURES

The Context

Our high school offers three levels of calculus: College Board Advanced Placement AB and BC Calculus courses and an Applied Calculus course. BC Calculus, the school's highest level course, uses Thomas and Finney's (1996) *Calculus* and maintains high expectations for mathematical rigor and proof. While they are not prohibited from using the calculator, BC students are not explicitly taught to use it for calculating limits. The methods they learn are all analytic, including epsilon-delta proofs. In BC

Calculus, the idea of a limit is first defined in the following way: "Let $f(x)$ be defined on an open interval about x_0, except possibly at x_0 itself. If $f(x)$ gets arbitrarily close to L for all x sufficiently close to x_0, we say that f approaches the limit L as x approaches x_0" (Thomas & Finney, 1996, p. 55). BC Calculus students are also given a formal epsilon-delta definition. They spend about two weeks very early in the school year explicitly devoted to the concept of limit. Those two weeks consist of just a couple of days on algebraic techniques with the remainder devoted to work with epsilon-delta proofs.

AB Calculus, the school's middle level course, uses the text *Calculus: Graphical, Numerical, Algebraic* (Finney, Thomas, Demana, & Waits, 1995). This course includes analytic methods for finding limits, but does not include epsilon-delta proofs. In addition to analytic methods, students in AB Calculus are taught how to find limits using tables and graphs on the TI-83 graphing calculator. In AB Calculus, limit is defined in the following way: "Given real numbers c and L, if the values of $f(x)$ of a function f approach or equal L as the values of x approach (but do not equal) c, we say that f has limit L as x approaches c" (Finney et al., 1995, p. 106). AB Calculus students spend about two weeks on the study of limit beginning in late September with the rehearsal of algebraic techniques accounting for the majority of this time.

Applied Calculus, the school's lowest level calculus course, uses the text *Contemporary Calculus Through Applications* (Bartkovich, Goebel, Graves, & Teague, 1996). Students in this class have constant access to the graphing calculator, but are introduced to limits in a laboratory style setting using a CAS, Mathcad. Intuitive approaches using tables and graphs are emphasized. Students learn basic algebraic methods for calculating limits, but there is no mention of epsilon-delta proofs. In Applied Calculus, limit is defined informally as "the target value that the output approaches as the input approaches the value listed." Applied Calculus students spend about two weeks in late October studying limit as a precursor to derivative. The two weeks consist primarily of experimentation with graphs and tables with only a couple of days devoted to algebraic techniques.

Participants

The participants were seven high school seniors in a small, reasonably affluent, university town who were currently enrolled in a calculus class. They were volunteers selected solely for their willingness to participate. I was not the instructor for any of these classes. The interviews took place in early November. All of the participating students had studied the concept of limit, but none had yet learned L'Hopital's rule for calculating limits.

All had constant access to the TI-83 graphing calculator and at least two years of experience in using it.

Two of the students, Carrie and Carl, were enrolled in BC Calculus. Both were identified by their teacher as average students in the class. Another two students, Ben and Bob, were enrolled in AB Calculus. Bob was identified as a high achiever in his class, while Ben was considered below average. The other three students, Alice, Abby, and Adam, were enrolled in Applied Calculus and were identified by their teacher as average students in their class.

Methods and Instruments

Each student was interviewed for approximately one hour. The interviews were audio- and videotaped for analysis. The interviews were designed around a series of tasks adapted from the research of others (Szydlik, 2000; Williams, 1989). My intention was to determine the methods students would use in various limit problem situations, the specific ways they would utilize technology, and the personal concept image that was held by each student. I asked students to "think aloud" as they completed tasks and I asked follow-up questions to clarify student methods, rationales, and thought processes.

Students were first asked to solve 10 limit problems aloud (see Figure 1). The limit problems were partially adapted from the research of Szydlik (2000) as well as problems from the Thomas and Finney (1996) calculus text. The problems from the Thomas and Finney text were well beyond where the participating BC Calculus students had studied in the text and there was nothing special about the nature of the problems that made them any more typical for any of the seven students. The first four problems referred to a graph of a function and were designed as a warm up for the students to start thinking about limit and for me to get some feel for how the subjects understood limit. My main focus was on the method students would use to solve the last six questions as all the participants had sufficient background to evaluate questions 5 and 6 algebraically, but none of them had the analytic skills to solve the final four questions without some use of technology. Further, questions 6, 7, 9, and 10 could potentially reinforce the conception of limit as unreachable, since each of the functions given in these problems contains either a hole or vertical asymptote at the limit point. Problem 8 could reinforce limit as a bound since the function approaches a horizontal asymptote as x approaches infinity, and Problems 7 and 10, might suggest that a limit is an approximation since the answers to these ($1/6$ and e) can only be approximated by the technology and the fact that there are exact values that the technology produced numbers approximate will often be unrecognized by the student. All students had access to a TI-83 graphing calculator during their interview session and the Applied Calculus subjects additionally had Mathcad available to them, as in their class.

The second part of the interview required students to respond to several statements from Williams's research (1989, 1991) indicating various conceptions of limit. First were the six true or false questions associated with the various identified limit conceptions described in Table 3.1. Next, students were asked which of these descriptions best represented their understanding of limit. Finally, in another attempt to determine each student's primary conception, I asked the students to describe in a few sentences what they understood a limit to be and, more specifically, what it means to

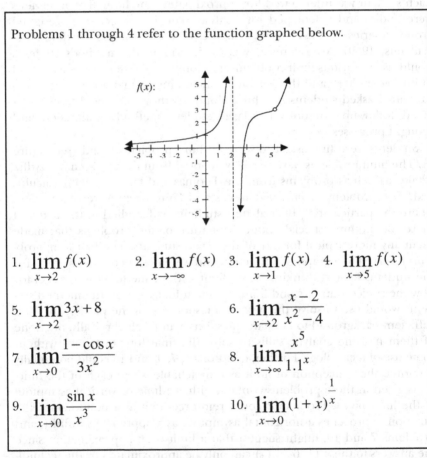

Problems 1 through 4 refer to the function graphed below.

$f(x)$:

1. $\lim_{x \to 2} f(x)$ 2. $\lim_{x \to -\infty} f(x)$ 3. $\lim_{x \to 1} f(x)$ 4. $\lim_{x \to 5} f(x)$

5. $\lim_{x \to 2} 3x + 8$ 6. $\lim_{x \to 2} \dfrac{x-2}{x^2-4}$

7. $\lim_{x \to 0} \dfrac{1-\cos x}{3x^2}$ 8. $\lim_{x \to \infty} \dfrac{x^5}{1.1^x}$

9. $\lim_{x \to 0} \dfrac{\sin x}{x^3}$ 10. $\lim_{x \to 0} (1+x)^{\frac{1}{x}}$

Figure 3.1.   Limit Interview Problems.

say that the limit of a function f as $x \rightarrow s$ is some number L. These statements were designed to pinpoint the student's personal limit concept so that I could draw connections between the students' solution techniques and their understanding of limit. For instance, would students who are prone to using the calculator's *trace* function be inclined to hold a dynamic view of limit, as Lauten et al. (1994) suggested?

Analysis

First, the techniques that each student used to evaluate each limit problem were recorded. Techniques included algebraic methods, graphing, graphing and using the calculator's trace function, graphing and evaluating the function at single points, generating tables of function values, and intuitive reasoning. Next, in order to assess each student's personal concept image of limit, responses to the true and false statements about the concept (see Table 1) were considered as well as the student's choice of a best description of limit among these. Each student's personal description of limit was also classified by its correspondence to the limit concept statements. These personal descriptions were coded according to the six views of limit identified in Table 1 and used to confirm the student's image of limit evidenced by the earlier questions. For example, students who described a limit as a particular value were coded as having a static view of limit. If a student described a limit as getting closer and closer to some point, then they were classified as having a dynamic view of limit.

RESULTS AND DISCUSSION

All students were successful on problems one through four (see Figure 3.1 for problems 1 through 10). Table 3.2 provides an overview of the solution methods used by each student for limit problems 5 through 10 and whether their process led them to a correct result. Table 3.3 lists each student's responses to the true and false limit concept questions and includes their choice of the conception that best suited them.

Table 3.2 shows that these students were able to find correct solutions using a variety of techniques. Students graphed and traced functions, used the calculator's "value" option to evaluate functions at particular values of x, and used tables to evaluate the functions and observe patterns. The real question for me was whether this ability to find correct solutions necessarily implied an accurate conception of the limit concept and, further, whether any of the techniques seemed to be associated with any particular conceptions.

Table 3.2.   Student Solution Methods

Participant	$\lim\limits_{x\to 2} 3x+8$	$\lim\limits_{x\to 2} \dfrac{x-2}{2x^2-4}$	$\lim\limits_{x\to 0} \dfrac{1-\cos x}{3x^2}$	$\lim\limits_{x\to \infty} \dfrac{x^5}{1.1^x}$	$\lim\limits_{x\to 0} \dfrac{\sin x}{x^3}$	$\lim\limits_{x\to 0}(1+x)^{\frac{1}{x}}$
Carrie	A	A	GT	GT	GT	GT
Carl	A	A	GE	I	G	GE
Ben	GE	GE	GE	G	GE	GE
Bob	A GE	A GE T	GE GT T	GE GT T	G	GE T
Alice	T	T	T	T	T	T
Abby	T	T	T	T	T	T
Adam	T	T G	G T	G T	T G	T G

Key:
 A: Used an algebraic technique, such as factoring or evaluating the func-
     tion at a specific point
 G: Graphed the expression on the calculator
GT: Graphed the expression on the calculator and then used the trace
     function
GE: Graphed the expression on the calculator and then evaluated the
     function at points close to the location of interest using the TI-83s
     "Value" function
 T: Generated a table
 I: Intuitive reasoning
Shading indicates failure to reach a correct solution
Multiple methods are listed in the order the student used them

It is clear that, at least in some cases, correct solutions to questions like those posed to these students do not insure an accurate conception of limit. For example, Ben was able to use informal techniques with technology to correctly answer nine of the ten questions, yet his concept image of limit was restricted to the procedures that he used to determine limits, giving him an understanding of limits as dynamic approximations that was strong enough to lead him to reject the formal, static definition of limit. As can be seen in the discussion that follows, other students also expressed ambiguous conceptions of limit.

Not every misconception about limit described in the research literature was exhibited by the students that I interviewed. While research suggests that students commonly view limits as boundaries, asymptotes, or points that cannot be crossed (Williams, 1989), none of the students that I interviewed exhibited this misconception (see Table 3.3). It has been suggested (Tall & Vinner, 1981) that expressions like "as x approaches s" might contribute to a view of limit as unreachable since it could be understood that x never actually reaches s, and while some of the functions that the students were given were undefined at the limit point, providing context for such a misconception to flourish, none of the students I interviewed expressed the belief that a limit is unreachable.

In the remainder of this section, I will illustrate the nature of the three most common misconceptions—Approximation, Dynamic-Theoretical, and Dynamic-Practical—and provide evidence of their connection with students' solution processes based upon statements they made. At the end of this section, I will provide some description of Bob's thinking, as he was the only student to both find all of the limits successfully and not reveal any misconceptions with respect to limit.

Table 3.3. Limit Conception Responses

Question	Participant						
	Carrie	Carl	Ben	Bob	Alice	Abby	Adam
1. Dynamic-Theoretical	T	T	F	F	T	T	T
2. Boundary	F	F	F	F	F	F	F
3. Formal	T	T	F	T	T	T	T
4. Unreachable	F	F	F	F	F	F	F
5. Approximation	T	T	T	F	T	T	T
6. Dynamic-Practical	T	F	T	F	T	T	T
Best Description	1	1	6	3	3	3	1
Subject Definition	1	1	6	3	1,3,6	3	1

Limit as an Approximation

Six of the seven students agreed that a limit is an approximation. For two of the students, Ben and Abby, there was clear evidence of this belief being rooted in their solution process. Ben generally chose to graph given functions and use the calculator's "value" option to evaluate given functions successively closer to the x-value of interest. For the problem $\lim_{x \to 0} (1 + x)^{\frac{1}{x}}$, Ben used the calculator to evaluate the function at x = 0, for which no value was given since the function is undefined at zero. He proceeded to evaluate the function at x = .001, obtaining y = 2.7181459 and at x = -.00001, which gave y = 2.7182954.

> Ben: My answer would be approximately 2.7182954
>
> Bill: And you know it's that because?
>
> Ben: That's a minimum value around, like it's close to zero and it's [the limit] just gonna be a little bit more or less than that value.

Ben's thinking is generally reasonable, though he did not recognize that "e" was the target of his approximation. He appeared to believe that there existed some exact value that he was approximating. However, additional probing revealed some weaknesses in Ben's understanding of the limit concept. I wanted to see whether Ben understood limit as a static value that the y values of a function could be made arbitrarily close to by restricting x values—as a formal definition would suggest. I asked Ben whether a limit is an approximation that can be made as accurate as you wish. Ben agreed, saying, "That's a step I go through." Limit as an approximation is a concept image of limit commonly held by calculus students (Davis & Vinner, 1986; Williams, 1989, 1991). Researchers have conjectured that this idea likely stems from expressions like "as x approaches s." Since students sometimes understand this as getting closer and closer to s, but at times not reaching s, a natural extension is that the limit is an approximation of the function at s. Abby, one of the Applied Calculus students who determined limits exclusively by generating tables, explained her agreement that limit is an approximation by referring to her solution process for $\lim_{s \to 2} 3x + 8$, "If I stopped at like $f(2.1)$ and 1.9 instead of going further, I might have guessed 13.9." For these students, the belief that limit is an approximation seems rooted in their solution process rather than in the language of limits.

Others seemed to choose limit as approximation simply to reflect their solution method. For example, Carrie stated, "if you knew how to do the algebra or like, morph the equation a certain way or if I know how to use the calculator the right way or something you could find a limit that's pretty accurate,…like, exact." She agreed that limit is an approximation because she believed she would otherwise be contradicting her own work since some of her responses were approximations. However, she seemed to be expressing that it was *her* solutions that were approximations while the actual limits would be precise values.

Students like Ben and Abby provide evidence that informal, technology-based approaches to finding limits can indeed contribute to the misconception of limits as approximations. This particular conception seemed clearly connected to the solution process for these students. On the other hand, while others expressed that limit is an approximation, for some, like Carrie, it is likely that this is only an expression for their limit solution processes and not necessarily their understanding of what a limit is.

Dynamic Conceptions

All the students except Bob also agreed with at least one dynamic conception of limit, and five agreed with both the dynamic theoretical and the dynamic practical conceptions. For some students, this seemed to be an incorrect belief that the limit really was a description of the movement of the function, while for others these notions were more part of their conception that enabled them to talk about how a limit might be determined. Consider Alice, who seemingly demonstrated a static view of limit, describing it as "a specific value," but also revealed a dynamic view (see quote below). Her explanations frequently contained a mix of dynamic and formal thinking. Alice's solution method was consistently the creation of tables using Mathcad. She accurately evaluated most limit problems using this approach of looking for a "stagnant" output with Mathcad set to provide answers to only three decimal places. Alice was satisfied that the limit had been reached when she could plug in numbers successively closer to the given x-value and reach an output value that did not change. For instance, when solving $\lim_{x \to 0} (1 + x)^{\frac{1}{x}}$, she concluded the limit was 2.718 because, "the closer I get to 0, the values [table outputs] remain at 2.718."

Alice's agreement with the dynamic limit conception of limit (see Table 3.3) seemed to stem from her solution process with Mathcad. For the dynamic-practical statement (see Table 3.1) she said, "that's what I did each time." Alice believed that the limit was a specific value, but that it could be reached at several x-values. "As x is approaching a certain point, the function is either remaining stagnant or it is changing. No matter what I do, once I get to a certain point, as soon as my x reaches a certain point, it [the function value] just stays at that point." She described limit as how the function moved. When the movement ceased, meaning the function became constant (in her eyes), she was convinced that the limit had been found. So while Alice believed a limit to be a static "target" value as defined in her Applied Calculus class, her conception is better described as an inaccurate dynamic conception. It also seems clear that her particular dynamic conception is a product of her solution procedure.

In the following exchange Carrie provides another example of the dynamic conception as she explains the concept of limit.

Bill: How would you explain to someone who doesn't know anything about limits, what a limit is?

Carrie: A limit is um, if you look at a function, it's the value that all the other values are approaching. Like if you keep on plugging in x-values, you'll keep on getting y-values that are gonna approach...Okay as all these x-values are approaching the x, then all the y-values are gonna approach $f(x)$. All the $f(x)$ values are gonna approach $f(x)$. Does that make sense?

Bill: What do you mean by $f(x)$?

Carrie: $f(x)$ is a function of x.

Bill: How does that relate to the limit?

Carrie: Okay, the limit is um, a certain value, like a certain, I guess value, as um, all other, like when you use x you plug it into a function and you'll get an answer which is $f(x)$. So as all x-values approach, let's say 1 or something, then um, all the $f(x)$ values for it, like each x-value that is around the x that you want, you plug 'em in and you'll get a y or y slash $f(x)$, whatever you want to call it, that's close to the $f(x)$ value that you would get if you plugged in the real number.

A little later when Carrie was asked what she would write as a definition of limit as x goes to s for some function, the following exchange ensued:

Carrie: It's as x-values are approaching this s-value, the closer they get, then they'll be approaching, then the $f(x)$ values are gonna be approaching the $f(s)$, that exact value, which is L [the limit].

Bill: What is L?

Carrie: $f(s)$, like pretend $f(s)$ is equal to L, if there wasn't holes or something like that. If L is $f(s)$ basically, like all the other, dang it, I can't say that 'cause there could be a hole there or something.

Okay, as *x*-values are approaching *s*-value, then they're gonna be approaching the number *L*.

Bill: What's gonna be approaching *L*?

Carrie: *f*(*x*) values are gonna be approaching *L*.

Carrie seemed to exhibit a formal understanding of limit as a static value, but explained limit verbally in dynamic-practical (see Table 3.1) terms. She described a limit as "the value that all the other values are approaching, like, if you keep on plugging in *x*-values, you'll keep on getting *y*-values that are gonna approach *f*(*x*) [the limit]" and added that limit is a "certain value." Carrie defined limit in what Williams (1989) calls dynamic-theoretical (see Table 3.1) terms, "As *x*-values are approaching this *s*-value, then *f*(*x*) values will be approaching *L*." Thus, Carrie understood limit in a way that is consistent with the formal definition, defined limit in a dynamic-theoretical way much as you might expect from expressions like "as *x* approaches infinity," and explained limit in dynamic-practical terms as one would work through a limit problem using a table or tracing a graph on a graphing calculator.

Formal Understanding

As the only student who was able to resist the notion of limit as an approximation or inappropriate dynamic conceptions, Bob deserves a closer look. The students that I met with did not have experience with L'Hopital's rule at the time of the interviews, thus could not use it to quickly determine the answer of ⅙ on the task $\lim\limits_{x \to 0} \frac{1 - \cos x}{3x^2}$. Seeing no algebraic way to evaluate this problem, Bob used his calculator to graph and explore the function using the calculator's "trace" option.

Bob: I want to see what's happening, like what the function's doing around *x* equals zero to see what I'm dealing with I guess.

Bill: Okay. So what are you thinking?

Bob: I just like to see what's going on…and all the y-values, how they're acting as I get closer to zero from both sides. I don't know. If I was just supposed to stop there, like if I was out of time I'd just write that it was one-sixth, on a test or something, but ah…

Bill: But what?

Bob: I don't feel like that's enough.

Bill: Why?

Bob: Because I can't support it algebraically in any way yet. So ah, I think that's the answer, but I'd like to be able to prove it. I don't feel comfortable; I don't know why it would be one-sixth. I feel there's a way that I should be able to get that somehow.

Bill: Do you have any doubt that it's one-sixth?

Bob: No.

Bill: But you said you don't feel comfortable that it's one-sixth.

Bob: Ya, I just, I don't know, I think that for every answer there has to be a way to get there without a calculator.

Bob clearly desired proof that informal approaches did not provide him. It was this need for proof that drove Bob to investigate the limit problems from different perspectives (see Table 3.2). Often, even when he was quite confident about a value he had obtained using one approach, Bob sought to verify his answer using multiple techniques in the absence of proof. He maintained a clear, static view of limit throughout the interview. He described the dynamic-practical conception (see Table 3.1) as a "method to find the limit, but that's not what the limit is." He described the process of finding limits in dynamic terms as his response to number two (see Figure 3.1), "because as x gets closer and closer to negative infinity, the y-value gets closer and closer to zero," but maintained that "there has to be some exact point." While Bob used approximating and dynamic techniques, examining graphs and tables, tracing, and evaluating specific points on the calculator, his concept of limit was clearly defined by the formal definition and was not negatively affected by any technology-driven solution processes. His own definition of limit is a combination of the static view he learned in class, and the dynamic-theoretical concept naturally developed out of phrases like "as x approaches zero." Bob stated that "a limit is an exact y-value that the function approaches or gets to as the x-values get closer and closer to s." This is the sort of firm, unwavering understanding that calculus teachers need to strive for and adequately question students to be certain that it is achieved.

IMPLICATIONS FOR INSTRUCTION

Four important implications can be drawn from this small, exploratory research project. First, right answers don't insure complete or even accurate understanding. It is clear that students can make productive use of technology to evaluate limits for which they do not posses the analytical tools to do so. However, my discussions with these students indicate that an ability to get right answers very well may not be an indication that they fully understand the concept. Students appear to commonly incorporate their processes in determining limits into their definition of the concept, so their use of technology has the potential to impact their understanding negatively as well as positively. Students who approximate solutions with technology may well believe that a limit is an approximation. Ben said that a limit is an approximation because approximating was a step he went through to determine a limit. Students who trace graphs or generate tables to determine limits may believe that a limit is dynamic; this was the case with Alice who said that a limit was dynamic because she determined limits by plugging in numbers to generate tables.

A second implication is that students need to be questioned deeply about limit to determine their conceptions, to reconcile potentially conflicting views, and to clarify their conceptions. My research supported that of others showing that students often appear to hold multiple conceptions of limit. The true nature of these conceptions can only be sorted out through deep questioning of students. Without deep questioning, I would not have been able to tell whether Carrie's understanding of limit was restricted to the dynamic procedures used, the idea of limit as an approximation as discussed earlier, or whether she did indeed understand limit more formally as a static entity, which the evidence suggests she did. I also wouldn't have been able to tell that Ben's correct answers were masking an inadequate conception of limit.

The importance of intertwining informal and formal approaches to limit is a third implication. As calculus courses continue to become more informal in approach in response to the availability of technology, educators need to be aware of potential implications of a strictly informal, intuitive style of pedagogy. Students with limited ability to use analytical methods for finding limits can make fruitful use of technology in evaluating limit problems. However, the intuitive processes for finding limits by examining and tracing graphs, evaluating points and examining tables do not necessarily lead to an accurate understanding of limit and can encourage some misconceptions or very limited conceptions. Fortunately, Bob provides an example of a student who did maintain an accurate personal definition of limit in a technological environment. Examining his thinking suggests the importance of instruction emphasizing a definition based on mathematical theory, even if informally expressed.

The fourth implication is that there is value in promoting a view of mathematics as making sense and provable. Bob not only maintained an accurate personal definition of limit, but this accurate personal definition was coupled with a desire for proof and understanding with respect to individual limit problem solutions. This desire for proof drove Bob to explore functions from multiple perspectives (see Table 2), seeking confirmation of results where proof was unavailable. Without such convictions, students may place too much faith in the technology that they are using and fail to recognize errors in output due to limitations of the technology. This points to the importance of designing instruction to generate beliefs that mathematics should make sense and be provable and to require that results be justified.

CONCLUSION

The widespread availability of technology is bringing the reform movement to life. Informal mathematical procedures and intuitive approaches to concept development are gaining favor with educators as they become more comfortable with the technology that often enables such approaches. Students can use a variety of informal techniques with the help of technology to evaluate limit problems that might otherwise be beyond their grasp. These intuitive approaches, however, may contribute to misconceptions about the limit concept, such as the belief that a limit is an approximation or that a limit describes the movement of a function around some target point as in dynamic conceptions. As we introduce technology and informal methods of solution to our classrooms, we must take care to ensure that students successfully integrate their informal knowledge of limit with the formal, static definition of limit.

The benefits of using technology may not be fully realized without careful examination of the relative risks that the technology itself brings to the learning environment. Thorough consideration must be given to the processes that students will use with the particular technology and what those processes add to the development of the concept. It is clear that technology can provide students with solutions to limit problems for which they might otherwise lack the analytical tools. However, it appears that when we incorporate technology toward these ends, the development of formal definitions and proof techniques remain important if misconceptions are to be avoided. Further, it seems that we can not rely upon students' abilities to calculate solutions to limit problems as the sole measure of their understanding of the concept. Now we must consider how we might best incorporate informal, technology-based approaches while maintaining the formality that is clearly needed. We must consider how we can best bring to light and sort out students' true conceptions of the limit concept. These are important issues that any of us might explore in our own classrooms.

REFERENCES

Bartkovich, K., Goebel, J., Graves, J., Teague, D. (1996). *Contemporary calculus through applications*. Dedham, MA: Janson.

Craighead, R., & Fleck, C. (1997). Calculating images: An experiment in teaching precalculus. *PRIMUS, 7*(4), 297-307.

Davis, R., & Vinner, S. (1986). The notion of limit: Some seemingly unavoidable misconception stages. *Journal of Mathematical Behavior, 5*(3), 281-303.

Finney, R., Thomas, G., Demana, F., & Waits, B. (1995). *Calculus: Graphical, numerical, algebraic*. Reading, MA: Addison-Wesley.

Gass, F. (1992). Limits via graphing technology. *PRIMUS, 2*(1), 9-15.

Lauten, A., Graham, K., & Ferrini-Mundy, J. (1994). Student understanding of basic calculus concepts: Interaction with the graphics calculator. *Journal of Mathematical Behavior, 13*, 225-237.

Monaghan, H. (1991). Problems with the language of limits. *For the Learning of Mathematics, 11*(3), 20-24.

Szydlik, J. (2000). Mathematical beliefs and conceptual understanding of the limit of a function. *Journal for Research in Mathematics Education, 31*(3), 258-276.

Tall, D., & Vinner, S. (1981). Concept image and concept definition in mathematics with particular reference to limits and continuity. *Educational Studies in Mathematics, 12*(2), 151-169.

Thomas, G., & Finney, R. (1996). *Calculus* (9th ed.). Reading, MA: Addison-Wesley.

Williams, S. (1991). Models of limit held by college calculus students. *Journal for Research in Mathematics Education, 22*(3), 219-236.

Williams, S. (1989). Understanding of the limit concept in college calculus students. Unpublished doctoral dissertation, University of Wisconsin-Madison.

CHAPTER 4

USING RESEARCH TO ANALYZE, INFORM, AND ASSESS CHANGES IN INSTRUCTION

Heather J. Robinson
Grayson High School

My fifth year of teaching proved to be the most interesting and eye opening for me to date. During this year I experienced many professional changes. At the beginning of the year, I became a "new" teacher again as I voluntarily transferred to a different school. Though my new school was in the same school district, changing locations was like entering a new world. The school's motto is "First Comes Learning," and I saw right from the start that every person there was dedicated to that vision. I was particularly impressed with the intentional scheduling of time for students to seek extra help from teachers. For example, each student has an advisory period during which they report to a homeroom location for attendance purposes. During this time students can request a pass to meet with a specific teacher. The school also has a rule that athletic practices and club meetings cannot begin until 45 minutes after the school day ends to allow students additional time to meet with teachers and work on homework. Different

Teachers Engaged in Research
Inquiry Into Mathematics Classrooms, Grades 9–12, pages 59–74
Copyright © 2006 by Information Age Publishing

disciplines are referred to as teams rather than departments and the team concept is prevalent. The support from administration is phenomenal in that teachers' concerns and desires are listened to, considered, and often utilized for school improvement. Teachers are encouraged to continue their education and participate in staff development opportunities in order to expand their expertise. All of these things were very different from my previous teaching environments. I was and continue to be extremely excited about teaching at this particular school.

At the same time that I was starting at this new school, I decided to enter graduate school. I loved everything about teacher education and found myself becoming highly interested in learning more about pedagogy and instruction. Also, as my teaching career progressed, many important issues came to the forefront in education, such as the No Child Left Behind Act, the scrutiny and debate regarding testing (particularly in the state where I teach), and funding, or the lack thereof, in education. I was eager to participate in discussions about these topics as a graduate student and to hear about the opinions and experiences of my professors and the other teachers in my classes.

Other factors that made this an opportune time of professional growth for me involved my school and district. During this time, my school district began offering many opportunities for staff development focused on research-based instructional practices. In addition, my school adopted a program called Learning Focused Schools. Learning Focused Schools (www.learningfocused.com) is designed to increase students' retention capacities by implementing researched based strategies.

With all these positive changes taking place in my professional life, I began to seriously consider changes I could make to become the kind of teacher that I had always wanted to be. My vision for my classroom has always been one where the students actively participate in the learning process by being engaged in enriching and meaningful learning activities that help make mathematics relevant and realistic—rather than an abstract "thing" out there to intimidate and use to separate the "haves" from the "have nots." I want my classroom to be a level playing field where students appreciate each other's ideas and are not afraid to be wrong in order to accomplish learning. Moving to a new school, combined with beginning a graduate program, provided the perfect opportunity to evaluate my teaching and make a purposeful effort to change. With four years of teaching experience, I had also developed some confidence in my ability to modify my teaching in a way that would positively affect my students' learning. Reflecting on my past four years, I had every reason to think that I had done a good job and that my students were learning; however, I still felt like I had not taken the time to focus on specific things that could improve my teaching and subsequently improve my students' learning. This chapter is the story of how I used research to better understand my teaching, to inform the changes I made, and to assess the results.

THE IMPORTANCE OF RESEARCH

One of the first courses I took in my graduate program focused on current research in mathematics education. I had never been truly engrossed in the research process and its practical application in mathematics education, but I quickly recognized that research is the key to efficient, effective change. After reading articles that described the changes that were brought about because of research, such as *Issues and Options in the Math Wars* (Schoen, Fey, Hirsch, & Coxford, 1999) and *Making Mathematics Work for All Children: Issues of Standards, Testing, and Equity* (Schoenfeld, 2002), I was convinced that research truly has a place in current educational practices and is crucial to effective teaching. Before this course, I viewed new instructional strategies as hit or miss—sometimes the lesson works well and sometimes it does not. From reading research journals, I have come to realize that research can inform the selection and implementation of instructional strategies and, as a result, make a huge difference in whether a strategy hits or misses.

Self-analysis and researching one's own practice seem particularly difficult for educators due to the isolating nature of our profession, with our only sources of feedback typically being the students we are charged to teach and an occasional administrator. Once we get over the feelings of resentment and resistance to change and open our minds to new ideas (something we often ask our students to do), we can reap the benefits of research that has been conducted in the past few decades and conduct our own research.

THE CONTEXT

The high school in which I teach serves approximately 2,500 students in a middle- to upper-middle class town. The area used to be predominantly Caucasian, but it is rapidly becoming more racially and culturally diverse. Nine percent of our student body receives free and reduced lunch. The student population is 82% Caucasian, 9% African American, 4% Hispanic, and 2% Multiracial. The mathematics instructional team consists of approximately 20 teachers. The students that I teach are all enrolled in a college preparatory curriculum, the curriculum taken by approximately 70% of our students. Our college preparatory sequence is Algebra I, Geometry, Algebra II, and then Trigonometry.

A staple course in my teaching career had been Algebra II, which I had taught each year of my career. When I received my course assignment for the first year at my new school, I was delighted to see three sections of Algebra II! Since the school is on a traditional six-class-a-day schedule, I was also teaching Advanced Placement Statistics and a general mathematics course titled Money Management. Because I had experience with and had developed a repertoire of lesson plans and activities for the Algebra II course, I thought this was the perfect course in which to focus on my teaching practices. My prior teaching experience with Algebra II had given me some insight into the topics that students have trouble understanding. Functions and simplifying rational expressions were the major topics that caused my students grief. In past years, I had gone through the process of rewriting some lesson plans from time to time, as most teachers do. Sometimes this "revamping" went very well, and my students responded by showing a better understanding than students in previous years. On other occasions, my revisions didn't improve student learning, but still gave me insights into aspects of my pedagogy and instruction that needed improvement.

Another reason I chose to focus on my teaching of Algebra II was that during my fourth year as an educator, our school district adopted a new textbook for this course (Schultz, Kennedy, Hollowell, & Ellis, 2001). This text included extensive ancillary materials that were specifically designed to enhance instruction, and consequently student understanding, by providing alternatives to instruction focused on the teacher presenting information. As I used this textbook during the first year of its implementation, I realized that I was not fully taking advantage of all the "extras" that were available to me and perhaps was depending on the textbook to provide the structure for my lessons, rather than using it to support my instruction. It occurred to me that I could be doing so much more for my students if I were to use my school district's curriculum as a guide for what topics were important, use the text to support instruction of these topics, and take advantage of the ideas that were provided to elaborate on these topics and develop deeper understanding. I began to take a closer look at my teaching style and all the components that made me "Mrs. R."

A PICTURE OF MY PRACTICE

I approached this revision of my teaching as a research project: I took a picture of my current teaching practice, identified the changes I wanted to make, made an effort to research and investigate the best ways to make these changes, implemented the changes, and assessed the results by comparing semester final exam scores. My first course of action was to develop an objective and honest picture of my teaching practices and habits. I videotaped myself teaching a chapter on functions. The reality was devastating: I discovered that I was a lecture-driven teacher who occasionally threw in a couple of activities. I was not teaching with a purpose—my lectures were intended to disseminate the required information and the activities were included more for variety than to extend my students' learning. My instructional practices were not engaging my students or encouraging them to actively participate in the lesson. Rather than carefully thinking through instruction and planning it to meet my goals, I read the textbook and then recited what I had read to my students. I saw evidence that the textbook served as my lesson planner rather than as a resource to help support the lesson.

Perhaps my motivation in depending so heavily on the textbook was to make sure that I "covered" the mandated curriculum in my school district, which is very specific for each subject. The district's goals for the students are very skill-oriented, and assessments are to be written to enable students to demonstrate that they have mastered the required skills. For example, one curriculum objective for Algebra II is that students thoroughly understand the concept and the mathematics of functions. The curriculum guide for Algebra II lists 25 requirements for students, including the following:

- Describe functional relationships.
- Identify, write, solve, and graph absolute value, step, and constant functions.
- Develop algorithms and analyze functions using the Fundamental Theorem of Algebra.
- Identify domain and range for algebraic and transcendental functions.

My students were able to perform mathematical operations with functions such as evaluating, adding, and multiplying functions, but I honestly was not sure that they ever understood what a function is and or how it might be applied. They knew all the rules for identifying a function, but had no idea what functional relationships imply. When they were asked questions that required deeper thinking, very few students were able to make the connections necessary to think about functions past the basic information that they had memorized. My role during a typical lesson about functions consisted of giving notes that included a definition (usually directly from the textbook), assigning some practice problems where students might evaluate a function or even create an x-y table and graph a function using a graphing utility, answering students' questions during the lecture or while they were practicing, and assigning the next night's homework. During the unit on functions I might include an activity or a discovery lesson, but, for the most part, conveying information through lecture was my main mode of instruction. My assessments for the unit included one or two quizzes and a chapter test at the end of the unit.

These assessments were either teacher-made, from the textbook ancillary materials, or a combination of the two.

By a certain set of criteria, my students were largely successful; they were able to do classwork and homework with very little trouble, and they were earning good grades on assessments. All in all, student performance indicated that I was doing a good job of helping them meet the expectations for success as defined by my district. However, there was one area where my students did not experience a comparable level of success—the final exam. The final exam for each semester was a comprehensive, multiple-choice examination that was comprised of questions selected from an electronic item bank maintained by the district. Each school was responsible for selecting exam questions from this item bank and creating a subject area exam that covered the required curriculum, and every teacher of a given subject was required to give the same final exam. Final exam scores were significantly lower than students' scores from their prior teacher-made assessments for all Algebra II classes, but I was particularly disturbed by the fact that my top students often earned failing or near failing grades on the final exam. For example, Kim, a student in my Algebra II course during the 2001–2002 school year, had a 96% course average prior to the spring semester final exam, but her final exam score was 44%. This, in addition to the other factors mentioned previously, became motivation to examine the effectiveness of my instructional practices.

It would have been very easy for me to continue teaching in my new school as I had in my previous school, to keep my graduate school experiences separate from my classroom practices, to continue to use my old Algebra II lesson plans, and to attribute the final exam scores to student apathy or something else outside of my control. After all, other than those final exam scores, my students were showing every indication of success. Instead, I thought about what someone once told me was the definition of insanity—continuing to do the same thing and expecting different results. I had to ask myself: If I expected different results, what was I going to change to make that happen? It would have been a disservice to myself, my students, and ultimately to the teaching profession if I did not recognize my role in all these events and empower myself to implement change.

MAKING THE CHANGES

To ease my transition, I decided to develop an "action plan." I made a "laundry list" of things that I wanted to change about my teaching and how I planned to go about changing them. These changes included lecturing less, providing opportunities for higher-level student thinking, and developing classrooms norms to support student-student dialogue. In the sections that follow, I describe these changes, along with the actions I took to achieve them. I will conclude the chapter with an analysis of my students' performance.

Less Lecture!

According to Bloom's taxonomy (Bloom, 1984), memorization is the lowest level of understanding. Unfortunately, it is also the kind of learning likely to occur through lecture, the most typical type of instruction in U.S. secondary school mathematics classrooms (Weiss, 1994). By lecture, I refer to an instructional setting that focuses on the teacher and his or her knowledge and gives students information with little opportunity to contribute and minimal need to think. The first time I read "A Vision for School Mathematics" in the opening section of the National Council of Teachers of Mathematics (NCTM) *Principles and Standards for School Mathematics* (2000) was in my first graduate school course. As soon as I read it, I knew that this was, in fact, the vision that I had for my own classroom. I think the vision of NCTM implies that students learn best when they are actively engaged in thinking about and doing mathematics. In contrast, lecturing often curtails students' freedom and opportunities to think. Often, they are so busy trying to write down every word that is said, that they are missing the big mathematical ideas. In addition, my lectures were very skill-focused and not concept-focused. One student wrote in her Math Autobiography, "Math teachers just tell you stuff. It's like 'learn this skill so that you can get it right on the test.' Nobody ever tells you why you learn it, you just have to learn it." That's not the image of mathematics or of learning that I wanted my students to have.

One of my goals was to move toward a learning environment in which students would be motivated to learn. Kazemi and Stipek (2002), in a study of student motivation in classrooms where teachers were teaching for understanding, found that "orienting students toward learning rather than toward performance engenders more active learning strategies" (p. 22). Creating a student-centered environment that focused on learning was a colossal challenge for me. To do so, I had to build a framework that I found manageable and that allowed me the time and opportunity to plan for and implement these new strategies.

I began by putting myself on a time table that restricted my lecture time to a maximum of 60 minutes per week. I defined "lecture time" as time spent presenting a skill or terminology to the students through rote demonstration. By limiting my lecture time, I was able to use time I would normally have spent writing lecture notes to plan new strategies and to think about the logistics of implementing them.

The Learning Focused Schools model provided a multitude of strategies for motivating students. Two that I implemented on a daily basis were: (a) beginning with an essential question that provided focus for the lesson, and (b) instituting activating strategies to get students interested and gear their minds towards answering that essential question. For example, an essential question that I formulated to open the chapter on functions is "What is a function and how can I identify it?"

The textbook ancillary materials were a major source of ideas for alternative teaching strategies and problems; I used both the lesson activities and cooperative group activities frequently and at different points in my lessons. Another valuable resource was the book *Worksheets Don't Grow Dendrites* by Marcia Tate (2003). In this book, Dr. Tate lists 20 brain-based thinking strategies for increasing students' learning. She explains how to make instruction more student-centered by teaching in ways that have been researched and shown to help students maximize their learning potential.

When I did lecture, I involved students in the lecture by using guided practice, giving them scenarios or problems to work on individually or in groups, and asking them questions such as, "Why do you think this looks this way?" or "What will happen if we change this value?" This new form of lecture usually resulted in students talking more than I did!

Including Opportunities for Higher-Level Thinking

In my undergraduate program, I took an entire course dedicated to Bloom's taxonomy (Bloom, 1984) and writing questions at each level of the taxonomy, yet, for some reason, I completely abandoned Bloom's taxonomy as a practicing teacher. As part of my graduate program, I was asked to compare an assessment that I had written to Bloom's taxonomy and analyze the level(s) of questions I was asking my students. I chose to use what I considered one of my best quizzes (Quiz 6, shown in Figure 4.1), which tested knowledge of exponential functions.

This quiz consisted of 10 procedural questions that require students to use only application level competence. I was ashamed by the lack of analysis, evaluation, and synthesis level questions on my quiz! After this exercise, I revisited the quiz to incorporate questioning at more levels of the taxonomy. Many of these questions came from the Enrichment Masters and Alternative Assessments included in the textbook ancillary materials. For example, I added the question in Figure 2 to incorporate analysis level questions. In this question students are not only asked to "do the math," but are required to defend their responses with facts and drawings—demonstrating that they can fit all the pieces together.

QUIZ 6

Tell whether each function represents exponential growth or exponential decay.

1. $f(x) = 5(1.2)^x$

2. $f(x) = 10(0.8)^x$

3. Sam has a choice between an investment that pays 6% annual interest compounded monthly and an investment that pays 5.9% annual interest compounded daily. Which investment will earn Sam more money over a 10-year period of time if he invests $1,000?

4. Write the equation in logarithmic form: $9^{1/2} = 3$

5. Write the equation in exponential form: $\log_2 64 = 6$

Evaluate each expression.

6. $\log_{12} 12^8$

7. $9^{\log 32}$

Solve each equation for x.

8. $\log_x 216 = 3$

9. $\log_8 (3x - 7) = \log_8 (48 - 8x)$

10. $\log_2 (2x) + 2 \log_2 (x - 6) - \log_2 (x + 4)$

Figure 4.1.   Quiz Prior to Assessment Changes.

After giving this quiz to students, I became blatantly aware of my students' lack of critical thinking and communication skills and of the absence of an emphasis on these elements in my instruction. I gradually began infusing more critical thinking skills in my class activities and adding questions to homework and in-class assignments that required deeper thought. Instead of giving students a set of procedural questions, I gave them tasks that were a mixture of skills, reasoning, and communication. I also began to vary the types of questions on assignments and assessments—multiple-choice, short-answer, and open-ended; I made sure that students had to explain their reasoning on at least two questions. For example, I revised a test on functions from 20 multiple-choice questions to 12 multiple-choice questions and eight questions that were open-ended. My goal was to use the multiple-choice and

---

Use your knowledge of exponential functions to complete the following tasks:

a.   Graph $f(x) = 125\left(\dfrac{1}{5}\right)^x$.

b.   What is the domain and range of this function?

c.   Use your knowledge of exponential functions to convince me that f is an exponential function.

d.   Describe how you can determine whether f is an exponential growth or decay function.

e.   Compare the graphs of $f(x) = \left(\dfrac{2}{3}\right)^x$ and $g(x) = \left(\dfrac{3}{2}\right)^x$. Describe how you can use the graph of f to graph

   g. Support your description with a drawing.

f.   Graph $f(x) = (3)^{-x}$ and $g(x) = \log_{\frac{1}{3}} x$. Explain why the graphs are symmetrical with respect to the line y = x.

   Demonstrate your response with a drawing.

Figure 4.2.   Sample Question from Quiz After Assessment Changes.

open-ended questions to address all levels of Bloom's Taxonomy. The "before and after" test question in Figure 4.3 reflects changes in the ways I wanted my students to demonstrate their knowledge.

Over time, I was amazed at how well the students adapted to answering more demanding questions. As I changed the types of questions that I asked students, a parallel shift occurred in the types of questions they asked in class. Their questioning moved from "How do I...?" to "What happens if...?" or "Why does...happen?" It is important to note that my assessments are works in progress. As I improve my instructional techniques, I can expect more from my students. Engaging in this research process has heightened my awareness of the need to always strive to be a better teacher.

Original Question	Revised Question
Simplify $(14x^3y^5)^{-2}$	Joshua simplified the expression, $(14x^3y^5)^{-2}$. His final answer was $14xy^3$. The teacher marked Joshua's answer as incorrect. Name two things incorrect about Joshua's answer. Rework the problem to reveal the correct answer.

Figure 4.3.   Before and After Exam Question

Student-Student Mathematical Discussion

Developing classrooms norms that supported student-student discussion was by far the most difficult of my goals to accomplish. I was used to "controlling" what happened in my classroom, and I thought it was important for me to be "in control." To some degree, I could maintain this control while I was lecturing, so giving my students the opportunity to explore mathematics proved challenging for me. Prior to this research experience, I thought that a noisy classroom could not be a positive learning environment. If my students were talking to one another, I assumed they were off task. As I worked to change my teaching practice, I began to incorporate cooperative group activities into my lessons. Within these cooperative groups, every student had a responsibility to communicate with the other group members to make sure everyone in the group was accountable for their learning. They also completed a group synthesis evaluation in which they either wrote a report or communicated their findings to their classmates in another way.

Think-Pair-Share (TPS) was one cooperative learning tool that I used to achieve this small group interaction. In TPS, each student partners with another student and each pair has a question to answer or topic to address. They share their ideas or answers with their partner and are asked to share their conclusions with the class. Not every pair will look at a problem or mathematical scenario in the same way and this generates rich discussion among the whole class and small groups of students. The role of the teacher is that of the facilitator who provides the topic of discussion and raises questions to deepen the discussion.

One of the best opportunities for students to engage in mathematical discussion that was not directed by me came from the use of the Jigsaw approach to teach solving equations and conic sections (see www.jigsaw.org for more information). The students were divided into expert groups in which they learned a specific topic. Each "topic expert" was then expected to teach his or her topic to a new group that consisted of one expert representative from each of the original groups. The Jigsaw was quite extensive and spanned several class sessions, but the students were almost completely responsible for their learning. In my role of facilitator, I gave the students specific tasks to complete and provided feedback on their plans for sharing their knowledge with their classmates so that I could be sure the curriculum objectives were being met. The students' responses to this change of roles were awesome. Many of my students who had not previously spoke in class contributed to discussions and helped their peers. This also created an open forum for those students who didn't understand something to develop the faith and confidence to go to a peer and ask for help.

The power of peer tutoring was fascinating to me; I was amazed at how much students could learn from each other. One of things that made this so valuable for the students was that peer tutoring removed the intimidation factor that might have come from being put "on the spot" by me, the teacher—despite my attempts to put them at ease. In general, students are often more comfortable talking to each other than to the teacher because they may be fearful of saying the wrong thing even in the most inviting of learning environments. Small peer groups often provide students with an acceptance and a sense of cohesiveness that is invaluable. Getting students to this point was definitely a progression for me. Initially, they seemed resistant to voicing their own thoughts because of a fear that I would hear and scold them or correct them in front of the class. Once they realized that neither I nor their classmates were going to discredit their ideas and that there was mutual respect for all ideas, the discussion began to be more productive. When I took a participant's role rather than an authoritative role, my students began treating me as part of the discussion instead of as the discussion leader. They even began to question my ideas and require further explanations from me, rather than accept everything that I said as unquestionable truth.

My students now engage in rigorous and topic appropriate discussions on a daily basis. For instance, when introducing the topic of functions, I place students into cooperative groups and give them a set of functions that they have to graph, describe specific attributes of, and group in some way. Their findings are the starters for the next several class sessions where we classify parent functions and get into deeper discussions about the specific functions. This activity focuses on the students' thinking and promotes discussion of mathematics. The students have certain tasks to complete, but they must synthesize their findings and develop logical conclusions, rather than practice specific skills. I have observed that their findings are much more meaningful to them when they have been discovered rather than memorized. My instructional model has evolved to posing a carefully chosen problem, letting the students decide the best way to find a solution, and questioning them to deepen their understanding. As a result, multiple strategies routinely emerge, and the discussions often end with everyone—including me—learning something new. I look forward to finding out what my students will teach each other and me.

ASSESSING THE RESULTS

After reviewing videotapes of my teaching that were recorded prior to the changes that I made, it was clear that I had spent approximately 75% of any class period using teacher-centered strategies. The other 25% was spent on daily management tasks like checking homework or answering questions. Rarely would a student ask questions or submit an answer to a question that I had posed. After making the changes to my instruction, I spent approximately 30% of any given class period using teacher-directed strategies. I took on the role of discussion leader, in which I would introduce a new topic by building on prior knowledge from the previous day or previous topics. Directed guided practice began to elicit the "why"; students were much more interested in why they missed an answer or why they worked the problem a different way and got the same solution. Most of the class period consisted of students engaging in discussion about a topic or working on an investigation or activity either individually or in groups. More class time was allotted to students presenting findings to the class and defending these findings. As a result of consciously making these instructional decisions, my students flourished. I found that they were much happier, eager to come to class, and more ready to learn. I attribute this to my clear communication of a purpose for every class through essential questions and the students' knowledge that they would have the opportunity to actively participate in their learning.

Although I noted changes in my students' motivation and levels of understanding throughout the semester, I wanted to see if these changes had a measurable effect on their final exam scores. I began deliberately making changes in my teaching practice in January of 2003, the beginning of our second semester of the 2002-03 school year. Table 1 and Figure 4.4 show the final exam scores for Semester A (2002), prior to integrating research into my classroom, and two semesters afterwards, Semester B of the 2002-03 academic year, and the Semester A of the 2003-04 academic year.

Table 4.1. Final Exam Scores Before and After Instructional Changes

	Semester A (2002)		Semester B (2003)		Semester A (2003)	
	Course Average	Final Exam	Course Average	Final Exam	Course Average	Final Exam
A	6	0	10	4	11	4
B	26	2	33	14	36	22
C	12	2	16	12	19	16
D	13	4	5	10	8	20
F	7	56	1	24	7	19
# of Students	64	64	65	64	81	81

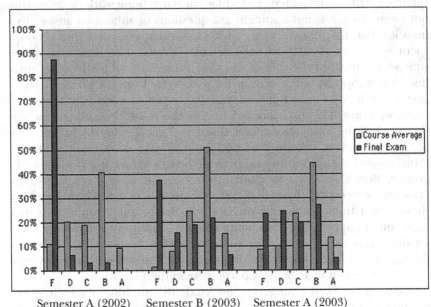

Figure 4.4.   Graphical Representation of Final Exam
             Scores Before and After Instructional Changes

When I began recording these data, the final exam grades were like a pyramid—there were a few students at the top with most students at the bottom. Ironically, the grades resembled the structure of my assessments— very few higher level questions with most questions being at the lowest level. They also reflected the tenuous mathematical understanding that my students had gained as a result of my skill-based instruction. As I made changes in my teaching and assessments, the exam grades became more evenly distributed. Of special note was the decline in the number of students who failed the final exam. These data convinced me that using a variety of instructional methods was helping my students be more successful. It is important to note that the format of the exam did not change over this period of time. The exam remained multiple-choice, primarily skill-based with a few conceptual questions. The student population also did not change in any noticeable way. Students were being more successful on the skill-based exam even though my instruction was no longer focused on repeated practice of isolated skills.

Although it is clear that my students were learning and retaining more than they were prior to my changes, I think I learned even more than they did. Examining my teaching has given me a new perspective on instruction. I have always thought of teaching as being similar to taking students across a bridge. It is my responsibility to get students from one side of this bridge to the other. Before this experience, they all had the same journey across the bridge—being carried by me without being allowed to "walk" on their own at any point. Now, I see each student's journey as an individual one and my job is to help him or her navigate to the other side. The navigation is much more powerful than the bridge itself.

My perception of my role as a teacher has changed dramatically through my engagement with research. In the past, I had underestimated the capabilities of my students. I believed that if I didn't explicitly deliver the curriculum, they wouldn't learn. I have come to believe that if students are encouraged and supported in using their own capabilities, not only will they learn, but they will also remember. I am convinced that learning is much more valuable when it is accomplished through active participation and engagement. As teachers it is our responsibility to provide students with the right tools to facilitate that learning. Our instructional practices determine the foundation for students' learning in the present and the future, and research holds the key for efficient, effective change.

REFERENCES

Bloom, B. S. (1984). *Taxonomy of educational objectives*. Boston: Allyn and Bacon.

Kazemi, E., & Stipek, D. (2002). Motivating students by teaching for understanding. In J. Sowder & B. Schappelle (Eds.), *Lessons learned from research* (pp. 17-22). Reston, VA: National Council of Teachers of Mathematics.

National Council of Teachers of Mathematics. (2000). *Principles and standards for school mathematics*. Reston, VA: Author.

Schoen, H. L., Fey. J. T., Hirsch, C. R., & Coxford, A. F. (1999). Issues and options in the math wars. *Phi Delta Kappan, 80*(6), 444-453.

Schoenfeld, A. H. (2002). Making mathematics work for all children: Issues of standards, testing, and equity. *Educational Researcher, 31*(1), 13-25.

Schultz, J., Kennedy, P. A., Hollowell, K. A., & Ellis, W., Jr. (2001). *Holt, Rinehart, Winston Algebra II Textbook Series*. Austin, TX: Holt, Rinehart, Winston.

Tate, M. (2003). *Worksheets don't grow dendrites*. Thousand Oaks, CA: Corwin Press.

Weiss, I. R. (1994). *A profile of science and mathematics education in the United States, 1993*. Chapel Hill, NC: Horizon Research.

CHAPTER 5

FROM TEACHERS' CONVERSATIONS TO STUDENTS' MATHEMATICAL COMMUNICATIONS

Florence Glanfield
University of Saskatchewan

Ann Oviatt
Deslisle Composite School

Darlene Bazcuk
Colonsay School

In 1990, teacher educator Virginia Richardson wrote a paper that offered responses to the following questions: "What is involved in bringing about significant and worthwhile change in teaching practices?" and "How can or should research aid in this process?" (p. 10). One of her suggestions was to create opportunities for teachers to "interact and have conversations around standards, theory, and classroom activities" (p. 16). This chapter provides a glimpse of what happened when one group of teachers, the

Teachers Engaged in Research
Inquiry Into Mathematics Classrooms, Grades 9–12, pages 75–96
Copyright © 2006 by Information Age Publishing
All rights of reproduction in any form reserved.

authors, had an opportunity to do just that. In the following we discuss our collaboration and the work that resulted from it.

GETTING STARTED

We are two teachers and a teacher educator who all have experience teaching secondary school mathematics in Canada: Ann and Darlene in rural schools and Florence in an urban school. At the time of our work together, Ann had 15 years teaching experience, Darlene had 27 years, and Florence had 20 years. Soon after Florence moved to Saskatchewan to begin teaching at the university, she had an opportunity to talk about assessment practices with superintendents from two local school divisions. She expressed her interest in working with high school mathematics teachers who shared her curiosity about students' understanding of mathematical language. After being contacted by their superintendents, both Ann and Darlene agreed to meet with Florence to discuss the possibility of collaborating.

Our First Meeting

Our first meeting was at the office of one of the school divisions for half a day about a month after the beginning of school. Immediately Ann and Darlene said to each other "I remember you! I've seen you at teacher's conferences." Reflecting on that first meeting, we laugh about how nervous we were: Ann and Darlene because this was their first experience working with someone from the university; Florence because she was new to the province.

As we introduced ourselves, we explained why we were interested in students' understanding of mathematical language. The initial explanations we offered were embedded in stories of our classroom and teaching experiences. A theme that emerged through these stories was the role of language in learning mathematics. We discussed the language a mathematics teacher uses while teaching a lesson—terms like "coefficient," "hypotenuse," "asymptote," "perfect square trinomial," "quotient," "variable," "equation," and "expression"—and how that language might help or hinder student understanding of mathematical concepts and procedures. We hypothesized what could be done to help students better come to understand this language.

Our interaction on that first day was what Clark (2001) calls "authentic conversation." Authentic conversation is an emergent phenomenon in which "the reconstitution of experience through personal narrative allows for safe exploration of uncharted territory and imagining the possible"

(Clark & Florio-Ruane, 2001, p. 12). Through our stories, or narratives, we were beginning to explore the use of language in learning mathematics. The intertwining of our conversation with stories of students in our classrooms allowed us to develop a shared sense of our experiences teaching high school mathematics; we could relate the students being talked about in others' stories to those from our own classes (Davis & Simmt, 2003; Glanfield, 2003). At the end of our first meeting we decided to meet one week later to continue our conversation and think about how, or if, our work would develop.

Our Second Meeting

We all remember Ann's first statement as she walked through the door for our second meeting: "I could hardly wait to talk more. All I've been thinking about and noticing in this last week is the language I use and the language my students use in mathematics classes." As we talked further, we discovered we had all implemented a mathematics curriculum with an explicit focus on the role of communication in learning and attended professional development workshops focused on communication strategies for the high school mathematics classroom.

The ministry of education in each province and territory in Canada is responsible for defining a "course of studies" or "curriculum" for schools. The curriculum describes the learning objectives for each course, and the philosophy behind them. The learning objectives include specific content students should learn and broad statements that are similar to the process standards of communication, problem solving, reasoning, connections, and representation that are described in the National Council of Teachers of Mathematics' (NCTM) *Principles and Standards for School Mathematics* (2000). One of the broad learning objectives in the Saskatchewan high school mathematics curriculum is that students should be able to "communicate mathematically" (Saskatchewan Education, 1996, p. 1).

We wondered what we might expect students to be doing in mathematics class if they are learning to communicate mathematically. We wondered about the role of the teacher in encouraging students to communicate mathematically. We wondered what a collection of activities that encouraged students to communicate mathematically would look like in a unit of study, or for a whole course. How would a teacher plan for, and implement, all of the different activities that were being recommended in the professional development workshops? Through these conversations we realized that if we were to work on this together we would have to focus our work on one unit of study within the curriculum—the concept of a whole course of study was too overwhelming.

We think back on this second meeting as the point where each of us decided that we were "hooked" and wanted to collaborate. During this meeting our conversation shifted from a concern with the language teachers use in the classroom to questions about engaging students in communicating about mathematics. We committed to continue our "wonderings" together, and to develop strategies that would encourage students to communicate mathematically. We agreed that before our third meeting, we would each consider the units of study in an 11th grade mathematics course that Ann and Darlene were both teaching in the second semester of the school year and identify a unit that would be suitable for encouraging students to communicate mathematically.

To facilitate our collaboration, Florence received a small research grant from her university. We used these monies to pay for substitute teacher and travel costs for Ann and Darlene, and to purchase audio and video tapes to record our work.

Our First Year

We discovered at our next meeting that each of us had identified the same unit, Complex Numbers and Quadratic Equations, as well suited for our work. Students tend to find complex numbers to be an abstract idea, and this unit's reliance on mathematical vocabulary reminded us of the stories we shared at our first meeting. For example, the term "simplify" is often used in the unit, and we had each seen students confuse "finding the solution" to an equation with "simplifying" an equation. It was also a unit where terms like "discriminant," "roots," and "x-intercepts," were used—terms that we had noticed some students avoiding.

During our first year of working together, we met approximately once a month. After our first two meetings at the school division office, we decided to meet at Florence's home to minimize distractions from our work. From our third meeting onwards, a pattern began to emerge in our conversations: we spent the first hour or so "catching up" by sharing stories of our teaching experiences since our last meeting and then shifted the conversation to "standards, theory, and classroom activities" (Richardson, 1990, p. 16). Once we decided that we would work together on a particular unit of study, our discussions focused on developing activities, anticipating how students might respond to them, and wondering what we might learn from those responses. After the implementation of the unit, the focus of our conversations shifted to analyzing the data collected and wondering what we might do next.

Our conversations were intertwined with not only our lived experiences as teachers, but with the theory that Florence was reading at the time. Florence shared ideas from her readings around communication and the use of language in learning mathematics as they applied to our conversations. These ideas grounded our discussions in the work others had done and helped us to see what we were experiencing in our classrooms as instances of something that existed beyond them.

Focusing Our Interests Through Readings and Conversations

We drew heavily from the NCTM *Standards* (1989, 1991, 2000) for the goals of our work together. In the area of communication, the *Standards* indicate that

> the development of student's power to use mathematics involves learning the signs, symbols, and terms of mathematics [...and that] as students communicate their ideas, they learn to clarify, refine, and consolidate their thinking.   [... Further, when] students have an opportunity to read, write, and discuss ideas in which they use the language of mathematics [that language becomes natural]. (NCTM, 1989, p. 6)

Mason (1988) focused our attention on how mathematical self-expression is an essential part of exploratory mathematical thinking; students' movement

> from a vague sense of something which is pre-verbal, to being able to capture that "sense of," involves taking time over trying to say what it is that I see, and then recording it in words, in mixtures of words and symbols, and finally in succinct symbols. (p. 49)

We were also struck by Mason's observation that "it is well worth spending time telling and re-telling someone" how we see what is going on (p. 20)—we wanted students to have that benefit rather than us.

Sierpinska (1998) helped us to think about how students need to negotiate meanings for terminology, not simply receive and repeat definitions:

> Meanings of mathematical expressions—words, formulas, diagrams—are found only when they become part of a discourse that the student shares with others. They are found when the student starts to use the new language and realizes that through its use, he or she can actually do something to and with others and achieve certain goals about which he or she cares. The meaning of a symbol is in the shared and actually used discourse as a whole. It is the discourse

as a whole that lends meaning to its parts, not the other way around. (p. 55)

Sierpinska also helped us to clarify our beliefs about the role of the teacher in a classroom dedicated to student discourse:

> ... transmission of knowledge is not an issue because knowledge is not in the head of the teacher. It is something that emerges from shared discursive practices that develop within the cultures of the classroom, the school institution, and the society at large. (p. 57)

The NCTM's *Professional Teaching Standards for School Mathematics* (1991) identified the teacher's role in discourse as one of six standards for teaching mathematics and elaborated on aspects of the teacher's role in discourse. These include:

> asking students to clarify and justify their ideas orally and in writing; deciding when and how to attach mathematical notation and language to students' ideas; and monitoring students' participation in discussions and deciding when and how to encourage each student to participate. (p. 35)

Figuring out what types of activities we could use to encourage conversation among our students about mathematics became an important feature of our work. From what we knew of our students, none of them, or very few of them had ever experienced a mathematics class where they were expected to clarify and justify their ideas or participate in discussions about mathematics. We discussed the ideas of "shared discursive practices" (Sierpinska, 1998) and "telling and re-telling" (Mason, 1988) someone about the ideas being generated and wondered if our mathematics classrooms encouraged students to become part of forming the discursive practices and whether or not there was a place in our classrooms for our students to tell and re-tell.

Through our discussions, we recognized that our classrooms did not provide a space for high school students to talk about mathematical concepts. Unfortunately, instances of the teacher "telling and re-telling" (Mason, 1988) were the norm rather than "discursive practices" (Sierpinska, 1998) generated through students' communicative activities. Our experiences of noticing the use of language in our classes, participating in NCTM *Standards*-based professional development, and digesting some of the literature, began to intertwine to frame our work. The many questions that kept emerging in our discussions started to overwhelm us, and we felt we needed to focus on connecting our understanding of the literature with our teaching and professional experiences. This reinforced our decision to develop activities for one unit specific to our teaching contexts and curriculum that would encourage our students to communicate about mathematical ideas. Not only would this benefit our students, it would also create an environment that would allow us to better investigate our questions about students' mathematical communications.

Even though we had a sense of direction, questions still emerged for each of us as we embarked on this professional journey: Will increasing the activities in one unit of study prevent us from covering the entire content in the curriculum? How will our students feel about the idea of having to communicate about mathematics when they likely have not done so previously? What is our role as teachers in our classrooms if students are doing all of the talking? It was with these questions echoing in our heads that we turned our attention to our research focus—what might we learn about students thinking about mathematics while they are engaged in activities designed to encourage them to communicate mathematically—and began to develop plans for our unit.

SETTING THE STAGE FOR STUDYING STUDENTS' MATHEMATICAL COMMUNICATIONS

We spent the approximately 5 months between our third meeting and teaching the Complex Numbers and Quadratic Equations unit developing a plan for implementing ideas that would encourage our students to communicate in our classrooms. This was a significant collaborative team project that both built on what we had learned from the work of others and created a space for our own research. In the following we describe the context within which we worked, our planning process, and the activities that we developed.

The Context

At the time of this study, Ann was teaching two semester-long 11th grade mathematics classes in a secondary school (7th through 12th grades) with a population of approximately 400 students, and Darlene was teaching one such class in a small K to 12th grade school with a total population of approximately 180 students. There were 30 students in each of Ann's classes and 16 students in Darlene's class. There were approximately equal numbers of boys and girls in each class and all but 2 students were of European ancestry. All students in the three classes had successfully completed the prerequisite courses, two semester-long grade 10 courses and one semester-long grade 11 course. The course that contains the Complex

Numbers and Quadratic Equations unit is elective in the sense that it is not required for high school graduation; on the other hand, it is required if students wish to attend college or university in the province. By enrolling in the course the students implied that they intended to pursue postsecondary education. In Ann's classes six students (approximately 10%) were repeating the course; in Darlene's class no students were repeating. All of the students lived in rural areas—in small towns, on farms, or on acreages. In Darlene's class all except for two students had been in the school since Kindergarten. Ann's school was a regional high school that drew students from several small towns and their surrounding areas. Both school communities contained a range of socioeconomic backgrounds.

Our Planning

Our first major decision was to plan the unit as an independent study. Students, working in self-selected groups of 3 or 4, would use the textbook together with other resources to explore the ideas in the Complex Numbers and Quadratic Equations unit of study. We had all heard about this idea at professional development experiences and chose the approach for two key reasons: (a) we wanted to shift the focus of the unit away from the teacher telling towards the teacher acting as an encourager for the students to dialogue among themselves (Smith, 1996); and (b) we believed that working independently with a group of self-selected peers would "create a context where students [could] safely express their own mathematical ideas" (Smith, 1996, p. 397). Our goal was to have students do the following:

(a) reflect upon and clarify their thinking about mathematical ideas and relationships; (b) formulate mathematical definitions and express generalizations discovered through investigations; (c) express mathematical ideas orally and in writing; (d) read written presentations of mathematics with understanding; and (e) ask clarifying and extending questions related to mathematics they have read or heard about (NCTM, 1989, p. 140).

We thought that the independent study would create safe opportunities for the students to engage in these activities with their peers. Further, we thought that this approach to the unit of study would provide students with the space and time to "tell and re-tell" (Mason, 1988) the ideas they were formulating and that "discursive practices" (Bauersfeld, 1988; Sierpinska, 1998) would be framed by the students.

Our second decision was to use the new textbook that Ann and Darlene's schools had purchased to use with the 11th grade mathematics class as a basis for the unit of study. We decided to do this because it provided a structure for our work and it was a resource available to all students. The textbook itself, however, had few (and in some lessons, no) opportunities for students to discuss mathematical ideas.

The activities we developed to supplement the text focused on encouraging students to interact in small groups, use the language of mathematics, and develop mathematics as a language. The mathematical content of the materials was based on making connections among mathematical concepts, previous experience and current experience. The activities were intentionally constructed in such a way that students, while working with peers, would have to think about, talk about, and then write about their understanding of mathematical ideas (Huinker & Laughlin, 1996). In the following sections we discuss the three activities that we developed: (a) the start-of unit project, (b) group assignments, and (c) the group presentation. These served as key data sources in our investigation of students' mathematical communications.

Start-of-Unit Project

Students completed a start-of-unit project in their groups. The idea of the project emerged in our conversations. We talked about how our students often do not seem to remember ideas they previously had studied. The intent of the project was to encourage students to make connections between mathematical concepts studied in previous courses and the new unit of study. We designed the project to require students to reflect upon their thinking about concepts previously studied and clarify their thinking about the relationships among these concepts (NCTM, 1989). The project also gave students a chance to express mathematical ideas in writing (by preparing their answers to the questions) and orally (by discussing the questions with their peers). We also hoped that working with their peers in preparing answers to the questions on the project would encourage them to participate in "questioning, listening, and summarizing" (NCTM, 1989, p. 140).

In the development of the project, we delved deeply into the new content presented in the Complex Numbers and Quadratic Equations unit and unpacked the assumptions the authors of the curriculum and the authors of the textbook were making about what students already understood. To facilitate these conversations, we each wrote down the assumptions we identified and explained our choices to one another. It was only when we all agreed that we began to create tasks. Sometimes the discussion was quite easy because we could look back at the curriculum for previous courses and know that students should have learned a particular topic. Other times the discussion was lengthier because there were "related ideas" that could not be clearly identified in a previous curriculum. In the end, the start-of-unit project consisted of 20 questions that were completed by groups of 3-4 students. Figure 1 (in the *Expressing Mathematical Ideas in Writing* section below) provides examples of questions from the start-of-unit project.

Group Assignments

The purpose of the group assignments was to have students reflect on and express their own understanding of the concepts that they had read about in the textbook. Further, the group assignments asked students to answer questions focused on the connections between and among mathematical concepts. A group assignment followed each of the six lessons in the textbook. Students were expected to work together in their group to complete the group assignment on a daily basis. The amount of time spent on the group assignment varied from group to group. Some of the groups completed both the group assignment and the lesson within the 60-minute period of time; others completed the group assignments in subsequent class periods. Figure 2 (in the *Expressing Mathematical Ideas in Writing* section below) provides an example of a task from a group assignment.

Group Presentation

At the completion of the unit of study, each group prepared a review sheet and a lesson for the class highlighting the important mathematical ideas within their assigned section of the unit. We wanted students to reflect on the mathematical ideas and hoped that the need to express the ideas orally would cause them to ask each other clarifying and extending questions about the material. One class (of approximately 60 minutes) was provided to students for the preparation of their presentation and one class was used for the six presentations. See the transcript included in the *Expressing Mathematical Ideas Orally* section for an example of the presentations.

Once the development of the activities and plans for the data collection were complete, Ann and Darlene began to implement the unit of study in their 11th grade mathematics classes.

THE PROCESS OF COLLECTING AND ANALYZING DATA

As students worked through the unit of study, we photocopied their written work from the activities we developed, audio taped their group conversations, and video taped their group presentations. Upon completion of the unit we also distributed a post-unit questionnaire that asked them about their experiences in the study.

Our data analysis took place after the implementation of the unit. We chose to do this for practical reasons—teaching the unit took place in the context of our regular working day and there simply were not opportunities for us to analyze the data as we were collecting it. Although all the students in the three 11th grade classes used the activities developed for our study, our data analysis included only the approximately 90% of those students for whom both the student and their parent/guardian agreed for their work to be used as part of the study. We further limited our data analysis to the 12 groups for whom we had a complete "set" of data: the start-of-unit project, all six of the group assignments, and the video taped presentation. (Due to the poor quality of the audio tapes, they were not included in our analysis.)

We set aside two Saturdays to review the students' work and to watch the video tapes. Although we met in the same place, we individually examined each of the 12 data sets. Our analysis of the students' work focused on two points: looking for evidence that our students were "reflecting ... clarifying ...formulating ... expressing ... reading and asking ... questions" (NCTM, 1989, p. 140); and observing what we could learn about our students' thinking from their responses to the tasks. During the two days of review, we drew on our experiences as mathematics teachers to keep field notes of our observations and explanations for those observations. We used these field notes as a way to keep track of what we noticed. Hodder (1998) suggests that "as the text is reread in different contexts it is given new meanings" (p. 111). Individually we provided different contexts as each of us brought our own meanings to the text we read and the videotapes we viewed. In the remainder of our meetings together that year we reflected on our explanations and meanings together and looked for themes in what we had individually noticed during those first two days of intensive data analysis.

WHAT WE LEARNED ABOUT STUDENTS' THINKING FROM THEIR MATHEMATICAL COMMUNICATIONS

In our work, we found that students were willing to engage in activities that asked them to "read about, speak about, reflect on, and demonstrate mathematical ideas" (NCTM, 1989, p. 140) in small groups. Through our analysis of their mathematical communications we gained some insight into the ways in which students think about mathematical ideas. In this section we share some of what we learned about our students as they expressed their mathematical ideas in writing and orally. We conclude the section with the students' perspectives on the experience. Throughout, we also share the effect of these learnings on us as the teachers and designers of the tasks.

Expressing Mathematical Ideas in Writing

The students in our study were willing to express their mathematical ideas in writing, but often used mathematical language in ways that made it difficult to determine their understanding of the concepts. This section discusses writing assignments that illustrate students' use of language and the issues that use raised for us as teachers.

In the study of complex numbers students are expected to know and understand the meaning of a real number and operations on the real numbers, topics they have studied in prior courses. In Figure 5.1, we give two examples of questions from the start-of-unit project that we used to assess the students' thinking about real numbers.

Consider the response of one group to the first task: "The definition of a real number is either a rational number or an irrational number. The set of real numbers R is the union of two disjoint sets. $\sqrt{-7}$ because there is no such thing as a negitive[1] root and a negitive plus a negitive is a positive." When we read this response we recognized that these students had copied the definition of a real number from the back of the textbook, as was the case for most of the groups, even though it was not our intent. The students' ideas in the statements "no such thing as a negitive root" and a

---

1. Here are examples of 9 different numbers:

$$-19 \qquad 0 \qquad \frac{1}{\sqrt{2}} \qquad -6.\overline{9}$$

$$\sqrt{-7} \qquad 1639 \qquad 0.04 \qquad -\sqrt{5}$$

$$\frac{2}{-3}$$

   a. What is the definition of a real number?
   b. Of the nine numbers above, which are not real numbers? Explain why.

2. a. Show how you would perform the following operation: $\sqrt{8} + \sqrt{18}$.

   b. Explain why we are able to perform the operation $\sqrt{8} + \sqrt{18}$ but that we are not able to perform the operation $\sqrt{2} + \sqrt{3}$.

Figure 5.1.   Sample Questions from Start-of-Unit Project. Reprinted with the permission of the Dr. Stirling McDowell Foundation for Research into Teaching.

"negitive plus a negitive is a positive" were mentioned by all the groups. These statements suggested that students were striving to make sense of the mathematics but raised questions about their level of understanding.

The same group's response to the second task was: "$\sqrt{8} + \sqrt{18} = 2\sqrt{2} + 3\sqrt{2} = 5\sqrt{2}$. The expression $\sqrt{8} + \sqrt{18}$ can be broken down with a common denominator but $\sqrt{2} + \sqrt{3}$ can't be added because they don't have a common denominator." Our intent in the second part of the first task was to see if students would connect the definition that they had written in the first part to the identification and explanation in the second part. We noticed that this group did not relate their explanation in the second part to their definition in the first part. These students could perform the operation related to $\sqrt{8} + \sqrt{18}$ and they related the task to finding a common denominator when adding fractions. All the groups were able to perform the operation $\sqrt{8} + \sqrt{18}$; other explanations offered for why the operation $\sqrt{2} + \sqrt{3}$ cannot be performed included: "...because when we break down $\sqrt{8} + \sqrt{18}$ the radicand is the same. In $\sqrt{2} + \sqrt{3}$ the radicands are different i.e. the radicand has to be the same" and "We cannot perform $\sqrt{2} + \sqrt{3}$ because the number under the root sign has to be the same when you break down." When we saw these responses we wondered if students knew why "the radicand" or "the number under the root sign" or the "denominator" had to be the same, and if they realized the differences as well as similarities between radicals and fractions. We found this generalization of language from prior experiences to the present setting very interesting. The task exposed misconceptions that we might not have otherwise been aware of and gave us ideas about understandings that students brought to the unit that could be built on in future instruction.

A second example to illustrate students' expression of their mathematical ideas in writing comes from a group assignment designed to provide the students with opportunities to reflect upon the mathematical relationship between real and complex numbers. Our analysis of the students' work provided additional insight into their thinking about this relationship.

Consider, for example, the question and group responses in Figure 5.2. The wording of the task was deliberately chosen to require more than copying a definition out of the textbook. When we examined the responses we saw students beginning to write about how they were seeing the relationship between a complex number and a real number. We saw evidence in both responses that students recognized the way in which a real number could be represented as a complex number, "because it can be written with an imaginary #" or that "you can always put $0i$ behind." The two responses also illustrate that students are attempting to explain why every complex number is not a real number: "...the imaginary part is not considered a mathematical 'real' number. It can not be used in the real world," and "complex #'s are not always real because sometimes you can't take the imaginary # out."

The explanation the first group provided for the second part of the question illustrates that they have some notion of "taking out" an imaginary

---

Explain why every real number is also a complex number but not every complex number is a real number.

Response 1:

③ Every real # is a complex # because it can be written with an imaginary #. (ex 6+0i, 9+0i) Complex #s are not ~~only~~ always real #s, because sometimes you can't take the imaginary # out.

Response 2:

Every real number is a complex number because you can always put 0i behind. for example 16 can be written as 16+0i. Every complex number is not a real number because the imaginary part is not considered a mathematical "real" number. It can not be used in the real world. For example $\sqrt{-25} = \sqrt{25} = 5i$. Although this is ~~not very~~ most likely true, the "i" can not be considered real.

Figure 5.2.   Example of a Group Assignment and Corresponding Group Responses Reprinted with the permission of the Dr. Stirling McDowell Foundation for Research into Teaching.

number. We interpreted this statement to mean that if you can't "take out" the imaginary number then the complex number is not real. What isn't clear in this response is what the students meant by the phrase "take out." We were reminded of the number of times we'd heard students say that they had "taken out" the common factor when factoring an expression and wondered if the students saw the structure of a complex number similar to the structure of a polynomial expression.

The explanation provided by the second group is attempting to connect the definition of a "real" number to their "real" world. What isn't clear is how students in the second group would think about a number such as 1.66666... Is this a number that could be "used" in the "real" world? We wondered where they might "use" it.

Although we noticed the ways in which the students understood (or misunderstood) the relationship between a real and a complex number, we were stymied as to how to address it at that point in time because we were reading through students' work together after the unit of study had been taught. Students were already studying the next unit by the time we could work together to carefully analyze their work. Our solution was to revisit the relationship between real and complex numbers during the end-of-year review by paraphrasing student responses to the questions in the Complex Numbers and Quadratic Equations unit, providing more formal definitions, and examining relationships between them through whole-class discussion. Upon reflection, we regret that there wasn't a time during the implementation of the unit for us to adjust our plans to build upon what we learned from an in depth analysis of the students' work.

Expressing Mathematical Ideas Orally

We provide the transcript of one portion of a group presentation at the end of the unit as an example of students expressing mathematical ideas orally and asking clarifying and extending questions related to the mathematical ideas being discussed. In this transcript Jason and Scott are a part of a group of four students. Jason begins his group's presentation about the section that introduced complex numbers:

Jason: Our group is doing section two point one and what I am going to talk about is the i's pattern first.

Jason begins to write and talks as he is writing "So, we know that i to the zero equals one, right? And i equals i, right? And i squared equals negative one and i to the cubed equals negative i, right? So, now what if you want to find i to the fourth?"

Class: Giggles.

Jason points to the list of equations that he's written on the blackboard and says "you have to understand that this is a pattern." He moves to the right of "$i^0 = 1$" on the board, writes, "$i^4 =$" and says, "so, we have i to the fourth…"

Girl1:   Jason, just wait. So, what is the pattern? So what, the pattern always goes….

Jason interrupts his classmate and says, "The pattern goes zero, one, two, three …"

Girl 1:   Okay guys, could you just explain this pattern one more time? I don't quite get it …

Jason continues to write and says "Okay, so just, so can anyone tell me what i to the fourth is?

The class begins to chant the remaining values for "i, negative one, negative i, …"

Girl 3:   So, what if you have i to the one hundred six?

Jason scratches his head and says, "i to the one hundred six?"

Girl2:   Try to break it down. …

Jason turns back to the blackboard, hesitates, says "okay" And then writes "$i^{106} =$"

Ann: Take one hundred six and divide it by four because there is four in your pattern.

Jason: Yeah, right. Does anyone have a calculator?

Ann: Do it with long division because you need the remainder Jason.

Jason computes $4\sqrt{106}$ and concludes that the answer is 26.2 Jason's group chants "no."

Jason: Remainder 2 or do I add a zero? Jason writes ".2"

Ann: Nope. …

Girl2:   Isn't it point 5? You have to add a zero. You get twenty six point five.

Scott: So that's i to the two, negative one.

Girl2:   That's twenty six point five.

Scott: No, we are looking at the remainder.

Jason: Yeah, you have to add the zero and bring it down. So it is the same as i squared? … So, now that you have the remainder of two you know that that'd be i squared and i squared on the Jason pattern that'd be negative one. Jason writes "$i^2 = -1$"

Girl 1:   How'd you get that? You have to look at the remainder?

Jason: Yeah, you have to look at the remainder.

Girl 1:   So, what do you do with the twenty six?

Jason: The twenty six? You just say "good bye." Jason uses his finger to wipe a finger line through the "26" on the blackboard, and continues "okay, anything else? Okay, I have one more thing to say about two point one and it is off to Scott."

Jason was primarily reading from notes during the first part of this transcript, and he constantly checked with his group to ensure that he was saying and writing what they had planned. Not only was he reading; it appears that they only focused on the pattern, not on how the pattern could be determined. We think that this transcript becomes interesting when a student from the class asks the question, "So, what if you have i to the one hundred six?" Jason now must try to find a way to explain how you would compute the value of "i to the one hundred six." This question prompted Jason and his group to consider their explanation. However, before Jason could start to offer an explanation, Ann gave Jason a method for determining the value of i^{106}. Notice how, following Ann's prompt for Jason, he and the students in the class shifted to negotiating the meaning of the result of one hundred six divided by four.

We chose this transcript for two reasons. The first reason was that we could see evidence that the students were willing to orally communicate mathematical ideas and negotiate with each other as they tried to "figure out" the value of i^{106}. Some students in the class were willing to ask clarifying questions about the mathematics that they'd heard about; for example, when Girl 1 says, "Jason, just wait. So, what is the pattern? So what, the pattern always goes…" Another student expressed an interest in extending the mathematical ideas that they'd heard about when he asked Jason, "So, what if you have i to the one hundred six?" Jason, although not certain about the way in which he could answer the question about the value of i^{106}, was willing to continue to be at the blackboard and communicate what he did know in mathematics.

The second reason that we chose this piece of transcript was because it highlighted one of the challenges that Ann and Darlene faced when implementing the unit that we'd planned. As Ann watched this videotape, she remembered how she had a hard time letting her students construct their answers. She said that "throughout the whole unit, I wanted to give students the answers. That's what my job is. That's what I'm used to doing as a teacher." It was especially hard for her as she watched the presentations in her classroom. She said:

I knew that I had never asked my students to make presentations before, so I felt somewhat responsible when they could not answer questions immediately. I always wanted to jump in with answers. In this presentation I jumped in without even realizing that I had jumped.

She continued:

it was only at the end of the class that I'd realized what I'd done with Jason's presentation. I had not given him the chance to show how he might've solved the problem. I immediately brought in my way of thinking about the pattern and the calculation and forced Jason to use my way of thinking.

As the three of us examined the video tape together, we realized that Ann's comments really pointed to our identity as high school mathematics teachers and the struggles we face as we implement new ideas (see Smith, 1996, and Clarke, 1997, for more discussion on this topic). We also wondered what Jason might have done if he "broke it down" in his own way. In general, Ann and Darlene both noticed that being quiet in order to listen to students' explanations (Davis, 1996) was one of their greatest challenges in our work together.

Student Perspectives

Student responses to the post-unit questionnaire suggest that opportunities to discuss the meaning of phrases and words within the context of mathematical tasks provided a basis for developing students' individual and group understanding of mathematical terms. As one student wrote, "It was good that we were in groups so that we could work together to help each other out. The activities did help us expand our mathematical vocabulary." A second student wrote that the experience "made me think and talk about the concepts more than I normally would." Another student eloquently described how interacting in groups helped individual students develop their own understanding of mathematical concepts: "Working in groups help us bring the ideas into form. We did a lot of different activities which allowed us to see things from different aspects."

The conversation was also critical to students' ability to begin to write about mathematical concepts. One student wrote, "I've never been able to write about mathematical concepts, but if anything helped, the discussion questions did, they forced you to explain it in writing." Another student valued the discussion questions on the group assignments, but offered,

"The discussion question would have worked if it didn't have a time limit, because of the added pressure, people would split the questions and do them theirselves. So there wasn't time for discussion, but sometimes arguments." We wondered what the student meant by the word "arguments" and to what extent the students' learning was affected by success of their group in using the tasks as they were intended. We regretted that we didn't have the audiotapes to help us better analyze the group interactions.

There was also evidence that students began to see the inter-relatedness of topics within the unit of study. One student described this in terms of the group assignments: "Group assignments also helped me see how each section in the unit fit together and what the whole unit was about."

CONCLUSIONS

Our work involved designing and investigating activities that promote and support students' mathematical communications in high school classrooms. Repeatedly policy documents invite teachers to engage students in the act of communicating with and about mathematical ideas. Because we suspected (and knew in some cases) that our students had not previously been given such opportunities, we designed and implemented activities for a unit of study that would allow us to examine what the communication of mathematical ideas might look like in our classrooms and for our students. The activities focused on providing opportunities for students to describe, discuss, and debate mathematical ideas with their peers. We also hoped that these activities would lead students to consider the symbolism of mathematics as a language that can be used to describe not only mathematical concepts, but also ideas in the real world. In broader terms, we wanted our students to engage in developing comprehension of the language of mathematics and mathematics as a language (Usiskin, 1996).

Our conversations and work together greatly expanded our sense of what was possible and how we might achieve it. Our understandings about teaching mathematics changed because we engaged in conversations about what we and others expect of students (standards), the theories we read, the classroom activities we developed, and the analysis of data we collected. As a result of our work, we understand better how to implement activities that encourage mathematical communication in our classrooms. We have also learned of some pitfalls we must consciously work to avoid as we go about changing our instructional practices. The habits we have developed within a perspective of mathematics teaching as a collection of procedures can lead us to do too much of the thinking for students even when we shift our lessons to focus on their understanding of mathematics concepts.

We will have the opportunity to use what we have learned and to encounter new challenges, we are sure, as we continue our research. We received funding from the Dr. Stirling McDowell Foundation for Research into Teaching to further explore the creation of activities to encourage student communication in the context of an entire course. We intend to address our question of how to both cover the content in the curriculum and encourage students to communicate mathematically. Based on our experiences in the first year of our collaboration, we are convinced that we can do this together far better than any one of us could on our own.

ACKNOWLEDGMENTS

This research was supported by the Dr. Stirling McDowell Foundation for Research into Teaching and the University of Saskatchewan President's Social Science and Humanities Research Council Fund.

NOTE

1. Quotes from the student work have not been corrected for spelling and usage errors.

REFERENCES

Bauersfeld, H. (1988). Interaction, construction, and knowledge: Alternative perspectives for mathematics education. In D. Grouws, T. Cooney, & D. Jones (Eds.), *Perspectives on research on effective mathematics teaching* (pp. 27-46). Reston, VA: National Council of Teachers of Mathematics.

Clark, C. M. (2001). Good conversation. In C. M. Clark (Ed.), *Talking shop: Authentic conversation and teacher learning* (pp. 172-182). New York: Teachers College Press.

Clark, C. M., & Florio-Ruane, S. (2001). Conversation as support for teaching in new ways. In C. M. Clark (Ed.), *Talking shop: Authentic conversation and teacher learning* (pp. 1-15). New York: Teachers College Press.

Clarke, D. M. (1997). The changing role of the mathematics teacher. *Journal for Research in Mathematics Education, 28*(2), 278-308.

Davis, B. (1996) *Teaching mathematics: Toward a sound alternative.* New York: Garland.

Davis, B., & Simmt, E. (2003). Complexity, cognition, and mathematics education research. *Journal for Research in Mathematics Education, 34*(2), 142-167.

Glanfield, F. (2003). *Mathematics teacher understanding as an emergent phenomenon.* Unpublished doctoral dissertation, University of Alberta, Edmonton.

Glanfield, F., Baczuk, D., & Oviatt, A. (2002). *Developing secondary students' mathematical understanding* (Research Report No. 63). Saskatoon, SK: Dr. Stirling McDowell Foundation for Research into Teaching, Inc.

Hodder, I. (1998). The interpretation of documents and material culture. In N. K. Denzin & Y. S. Lincoln (Eds.), *Collecting and interpreting qualitative materials* (pp. 110-129). Thousand Oaks, CA: Sage.

Huinker, D., & Laughlin, C. (1996). Talk your way into writing. In P. C. Elliott & M. J. Kenney (Eds.), *Communication in mathematics K-12 and beyond*. 1996 Yearbook of the National Council of Teachers of Mathematics (pp. 81–88). Reston, VA: National Council of Teachers of Mathematics.

Mason, J. H. (1988). *Learning and doing mathematics*. London: Macmillan Education Limited.

National Council of Teachers of Mathematics. (1989). *Curriculum and evaluation standards for school mathematics*. Reston, VA: Author.

National Council of Teachers of Mathematics. (1991). *Professional standards for teaching mathematics*. Reston, VA: Author.

National Council of Teachers of Mathematics. (2000). *Principles and standards for school mathematics*. Reston, VA: Author.

Richardson, V. (1990). Significant and worthwhile change in teaching practice. Educational Researcher, 19(7), 10-18.

Saskatchewan Education. (1996). *Mathematics A30, B30, C30 curriculum guide*. Regina, SK: Author.

Sierpinska, A. (1998). Three epistemologies, three views of classroom communication: Constructivism, sociocultural approaches, interactionism. In H. Steinbring, M. G. Bartolini Bussi, & A. Sierpinska (Eds.), *Language and communication in the mathematics classroom* (pp. 30–64). Reston, VA.: National Council of Teachers of Mathematics.

Smith, J. P., III. (1996). Efficacy and teaching mathematics by telling: A challenge for reform. *Journal for Research in Mathematics Education, 27*(2), 387-402.

Usiskin, Z. (1996). Mathematics as a language. In P. C. Elliott (Ed.), *Communication in mathematics K-12 and beyond*. 1996 Yearbook of the National Council of Teachers of Mathematics (pp. 231-243). Reston, VA: National Council of Teachers of Mathematics.

CHAPTER 6

LESSONS TEACHERS CAN LEARN ABOUT STUDENTS' MATHEMATICAL UNDERSTANDING THROUGH CONVERSATIONS WITH THEM ABOUT THEIR THINKING

Implications for Practice

Craig Huhn
Holt High School

Kellie Huhn
Holt High School

Peg Lamb
Director of NSF Bridges Project, Holt High School

For more than a decade, mathematics teachers at Holt High School (HHS)—influenced by the National Council of Teachers of Mathematics *Standards* (1989, 1991, 2000), the work of Oakes (1985) and Wheelock

Teachers Engaged in Research
Inquiry Into Mathematics Classrooms, Grades 9–12, pages 97–118
Copyright © 2006 by Information Age Publishing
All rights of reproduction in any form reserved.

(1992) regarding the impact of tracking students, and the inclusion movement of special education (Lipsky & Gartner, 1997)—have redesigned the school's mathematics curriculum to include all students. In particular, the Mathematical Sciences Education Board and NCTM's assertion that mathematical literacy is paramount for the workforce of today prompted members of the HHS mathematics department and school administrators to restructure the mathematics requirements such that every student is required to take Algebra I and Geometry.

Three years ago, mathematics, science, and special educators extended these restructuring efforts to promote scientific careers for all students, including those with disabilities, by developing a collaborative partnership with Lansing Community College (LCC). One component of the partnership involved a research team consisting of two HHS mathematics teachers (Craig and Kellie), two LCC mathematics teachers and a special educator at HHS with expertise in special education transition (Peg). From our experience working with students with disabilities, we knew that these students often test poorly and, as a result, their college mathematics placement is often below their level of mathematical knowledge. One thing we pondered was how to gage and truly understand the way and to what degree students, particularly those with disabilities, make sense of mathematics.

Based on the high school mathematics teachers' experiences interacting with students in their mathematics classes about their mathematical thinking and the insight gained from this type of discourse, we agreed that an interview process was the best way to acquire the kind of information we were seeking. This chapter focuses on two key questions that we investigated:

1. What do conversations with students about their mathematical thinking reveal about their mathematical understanding?
2. What pedagogical strategies might be indicated by what we learn about these students' mathematical understandings?

THE STUDENTS

The participants in our investigation were 12 students with disabilities from Holt and other area schools who were involved in the Bridges Transition Project, a project funded in 2000 by the National Science Foundation Division of Programs for People with Disabilities to develop a model for increasing the access of students with disabilities to careers in science, technology, engineering, and mathematics. These were students who, having interest in a career in these fields and planning to attend LCC, volunteered to participate in the Bridges Project. All 12 of the students had taken Algebra I, and 10 of the 12 had taken Algebra II. See Table 6.1 for more details about the students' specific disabilities and mathematics courses taken.

Table 6.1.   Demographics of Students Participating in NSF Bridges Math Research Project

Student (pseudonym)	Disability*	Math Courses Taken in their High Schools
Nathan	LD writing, ADD	Algebra I, Algebra II, Geometry, PreCalculus, Honors Discrete Math**
Muhammad	EI	Algebra I, Algebra II**
Adam	LD writing	Algebra I, Algebra II, Geometry, PreCalculus**
Trisha	LD writing, ADD	Algebra I, Algebra II, PreCalculus**
Elisha	HI	Algebra I, Algebra II, Geometry**
Nancy	LD in Rdg., Writing, Math Calculation, & ADD	Transitions Math, Algebra I, Algebra II**
Martin	EI	Algebra I, Algebra II
Jason	LD writing, ADD	Algebra I
Jessica	VI	Algebra I, Algebra II
Jeff	LD Rdg, Writing	Algebra I, Algebra II
Jamilla	LD, Rdg, Writing, Math Calculation	Algebra I, Algebra II
Jeremiah	LD writing, ADD	Algebra I, Geometry
Mason	LD reading, writing, math, S/LI	Consumer Math

*LD=Learning Disability; ADD= Attention Deficit Disorder; EI=Emotional Impairment; HI=Hearing Impaired; VI=Visually Impaired; S/LI=Speech/Language Impaired
**Function-based curriculum

As students with disabilities, they had very different experiences in their respective high schools. Six of the 12 students attended HHS: a suburban, full-inclusion 10th through 12th grade building that has developed a function-based, social constructivist mathematics program. The program's content focus on exploring relationships between changing quantities is coupled with the social constructivist belief that students construct knowledge based on interaction with their peers through groupwork and discussion. An inclusive high school is one that integrates special education students into the general education curriculum within regular classrooms, providing the support and accommodations required by their Individual Education Plan (IEP). Thus the special education students at HHS experienced the same mathematics reform curriculum as all of the regular education students at HHS. The remaining six students in the study were from other suburban high schools with varying beliefs about how to best teach mathematics and how to teach students with special needs. As a result, the classes that the students had access to, as well as the curricula for these classes, were very diverse.

THE PROBLEMS

Three problems were selected to assess students' thinking about particular topics in Algebra I content (see Figure 6.1). This choice was partially made to allow the same questions to be used for all students in the study, but also because we wanted to illicit some information about the students' foundational understandings of the main concepts of algebra. The purpose of the stock problem (problem #1) was to assess students' use of average rate of change to extrapolate in a contextual situation. The banking problem (problem #2), was included to assess students' ability to determine when two constant rate situations are equal. The "2x + 10" problem (problem #3), was used to determine whether students could differentiate between different symbolic representations commonly found in algebra. The problems were carefully worded so as to be non-biased to a particular strategy or method and to require a level of problem solving beyond a simple calculation. Our intent was to use questions that would present mathematics as "a sense-making discipline rather than one in which rules for working exercises are given by the teacher to be memorized and used by students" (NCTM, 2000, p. 334).

Problem 1:

Use the following table of stock prices to make an estimate of when the stock price will fall below $15.00 a share. Provide evidence.

Day	Price per Share
1	43.50
3	39.60
11	37.00

Problem 2:

You have $220 in a savings account, to which you add $30 every week. How long until you have $500 in savings? Your friend only started with $140 in savings, but puts away $45 each week. How long until your friend has more money in savings than you do? Explain.

Problem 3:

What does it mean to you when you see, "2x+10"? How about "2x+10=40"? Finally, what does it mean if you see "2x+10=3x+6"? Are these similar or different in any way?

Figure 6.1. Interview Problems

THE INTERVIEWS

The students were invited to participate in the interviews though a letter from the project director (Peg). The letter explained that during the interview they would be asked to solve two or three mathematics problems, share with the research team how they solved the problems, and then answer a few questions. Both the letter and our introduction at the beginning of each interview stressed that we were there to listen to what they knew, not what they didn't know.

The interviews were scheduled for a half hour each over the course of four days in the summer of 2002. Sessions were limited in time in order to schedule the interviews in advance. The setting for the interviews was an available room on the campus of LCC, which had video and audio recording equipment and a table that allowed the research team and student to sit together. All of the interviews were videotaped and each member of the research team took notes of students' responses and observed mathematical thinking. The interviews were scheduled with fifteen minutes between them, to allow some flexibility for the students to get to the interview site, and for the team to debrief on what we heard each student say and on the interview process. The videotapes, team notes, and student work samples were the data sources for this study.

Since this arrangement had the potential to be very uncomfortable and fraught with anxiety for the students, we were aware that we needed to be conscientious about creating a safe, non-judgmental environment. Given this concern, the research team attempted to make students comfortable by introducing themselves and asking each student about their high school mathematics classes and career goals.

Students were then given a problem to read (or have read if necessary) and think about. They had the option to work and then explain, or explain while they solved the problem. They were provided scrap paper and a graphing calculator. The first day we gave the problems in the order listed in Figure 6.1. For subsequent interviews we moved Problem 1 to the end because some students seemed to spend a disproportionate amount of time on it. The students generally had time to solve two of the three problems during the session.

During the interviews, we focused on gathering information about each student's degree of algebraic content understanding, command of this knowledge, ability to evaluate their own ideas and solution strategies, ability and willingness to use technology as a tool for solving problems, and level of confidence in their answer or solution. The three problems were the basis for gathering this information, but we were also aware that to accomplish our goal of gaining insight into the students' thinking we would need to ask carefully chosen follow-up questions. We focused on questions that built on students' actual ideas, not what we expected them to do; questions that were honest and sincere, showing a genuine interest in their thinking and a curiosity about their ideas, rather than being judgmental of their thinking. Examples of these questions include: What would that (number, symbol, or expression) tell you?; What does that stand for?; What were you thinking when you did that?; and Why did you change your mind?

We were also aware of the non-verbal cues that could distract from the purposes of the interviews. As teachers, we have a tendency to intervene assuming students are "stuck," thereby negating their opportunity to reason mathematically. We felt that it was very important for us to be patient, thus, during the interviews much time was spent in silence as the students thought. To gain insights into their conceptualizations, we asked probing questions without correcting their computational or problem solving errors. We were careful not to assert mathematical authority over their work, but instead to listen to their mathematical thinking, noting, but not verbalizing, the correctness of their answers and processes.

DATA ANALYSIS

During each of the clinical interviews, the members of the team took notes about their observations of the student's thinking and his or her mathematical process. Between interviews, the team debriefed about what they noticed, what they thought was interesting, and what content understanding they thought the student demonstrated. At the end of each day's session, we discussed in detail our findings from each of the student interviews, and arrived at a consensus about their mathematical understanding. After all of the interviews were conducted, team members viewed the videotapes for patterns and themes that were striking to us as researchers. Next, we met as a group to share the patterns and themes that emerged from viewing the videotapes and also verified our findings with our notes and the students' work. Finally, several of the interviews were transcribed as specific documentation to illustrate our findings. The vignettes in the following sections are narrations developed from the research team's notes, audio and video footage of the interviews, and conversations we had about our observations.

FINDINGS

The findings and the analysis will be reported in the context of vignettes describing elements of students' interviews, as is common with action research (see Hubbard & Power, 1999, particularly pp. 4–6, 15, 17–20). The findings presented in this paper are case studies born out of the conversations we had with the 12 interviewees about their understanding of the mathematics in the problems, and the behaviors we observed as they wrestled with their own view of their understanding. In looking at the data collected from these interviews, the research team identified some important issues related to the students' mathematical understanding that arose from our observations of the students as they worked and shared their ideas. These findings are organized into five main categories:

1. students' conceptualizations of "*x*";
2. difficulties with procedural knowledge;
3. the use of technology as a tool for learning;
4. the role of multiple solution strategies; and
5. students' experiences in learning mathematics.

After discussing these findings, we turn our attention to our conjectures about the pedagogical strategies that are indicated by them.

Students' Mathematical Understandings

Students' conceptualizations of "x." The bank problem (problem #2) highlighted differences in how students understand the meaning of "*x*." The following descriptions from student interviews illustrate their different conceptualizations of the idea of a "variable." Eight of the students worked on the banking problem. The panel noted that there was a clear distinction between two general types of answers given by the 8 participants. Three students (Muhammad, Nancy, and Nathan) were able to correctly answer the second part of the problem where they were asked when the two accounts would be equal (i.e. some time after the fifth week the "friend's" account had more then "your" account). Jessica's response illustrates the conceptual error made by the 5 students in this study who did not consider that the money in both accounts was changing over time at two different rates.

> After reading the first part of the bank problem, Jessica subtracted the $220 from the goal of $500, and divided the resulting $280 by $30 to get "about 9 weeks." When asked if she thought she was right, she was unsure, but decided to check. She stated that if you multiplied 30 dollars by 9 weeks, and then added that to $220, you should get about $500. She performed this operation, and was pleased to get $490, pointing out that it should be a little less since it didn't work out exact. When presented with the second part of the problem, she reiterated the problem, typical of students with problems in reading comprehension. She decided to subtract 220 from 140 to see how much of a difference there was to catch up. Knowing that the friend adds $45 a week, she confidently pronounced that it would take only 2 weeks before the $80 would be made up.

The four other students made the same conceptual error, although they approached the problem in different ways. These students simply worked to find out when the friend's account would reach 500 or more, answering slightly more than 8 weeks. These students apparently did not realize that the friend had more money in the account long before they reached the savings goal of $500. By contrast, Muhammad was one of the three students who were able to solve the second part of the problem, which demands more complex reasoning:

> As the first part of the problem was read, Muhammad muttered that its "gonna answer 500" and wrote "=500" down on his paper. He subtracted 220 from 500, and divided the answer by 30, reporting that the answer was "9 and a third weeks." When given the second part, he clarified, "is this starting at the same time?" Then, thinking out loud, said, "…need at least 501 or 500 point something… assuming that it takes the same amount of weeks… which shouldn't be done." He then checked to see how much the friend would have at 9 1/3 weeks, and finding $560, commented that it won't take 9 weeks. So he made a list on his paper comparing the amount in the bank for each account for each week. He labels "140 – 220" as week 0, and continued until he indicated week 5 with "5w 365 – 370" and week 6 with "6w 410 – 400." He said the answer was between week 5 and week 6 and that he could find it more specifically if it was necessary. It was clear to us from this statement that he would continue guessing and checking with values between 5 and 6 to be as specific as we desired.

The other two students who were able to get an accurate answer had similar understandings. Nancy evaluated both accounts at different weeks, answering that after 6 weeks, "they have more money than you by $10." Nathan put the function for each account into the graphing calculator and looked on the graph for when the two lines intersected, announcing that the accounts were equal after "5.3" weeks, where both are "at $380." Even though Muhammad, Nancy and Nathan used different strategies to solve the problem, all three recognized that the accounts changed at different rates and that the amount of money in each account changes over time.

One thing we noticed is that the three students who arrived at a correct answer attended HHS, where algebra is taught using a functional approach in which x is treated as a changing quantity rather than as a placeholder for an unknown number. The students who answered the second part of the banking problem incorrectly reduced the question to a problem where they found when the "friend's" account contained 500 dollars, errantly assuming that "their" account remained constant at 500 dollars, while the friend's account constantly increased. We wondered if this was more than mere coincidence and if our anecdotal evidence supported the idea that a more sophisticated understanding of algebra is attainable with a function-based perspective, as has been suggested by others (Chazan, 1993; Chazan, 1999; Heid, 1996; Heid, Choate, Sheets, & Zbiek, 1995; Yerushalmy & Gilead, 1997). Although our observations from this study are not necessarily generalizable to a larger population, we conjecture that the students who interpreted the second part of the banking problem as a static problem used the strategies they were most comfortable with from thinking of "x" as a placeholder for a specific unknown number and were unaware that they were missing the main point of the question.

With the conceptualization of "x" as a placeholder for an unknown number, instruction often focuses on solving for the value of "x" as the central skill that demonstrates students' "mastery" in algebra. Promotion of this understanding of "x" can diminish algebra to a guessing game, whereby being successful means having the best and fastest way of finding the one correct answer. Thus, this problem solving process is often taught with gimmicks or tricks so students don't make "silly mistakes." They are given many possible variations of problems, so they can practice "undoing" or "isolating the variable." In this type of methodology a student needs to practice many problems to be efficient and successful.

Each of these five students missed the concept that as the number of weeks gets larger, the initially lower account increases at a faster rate than the other account. Four of these students, when asked when the second account has more money than the first, immediately looked for how many weeks it took for the second account to also reach $500. They were apparently unaware that the two accounts could have equal amounts in them unrelated to a specific output like the $500. In this way, they were unable to think about the idea that the amount of money in both accounts is continuously varying, and instead tried to fix the output (using 500) in order to solve like they did in the first half of the problem. For example, Jessica thought the $80 difference in the starting amounts would diminish after just two weeks with the second account increasing $45 per week. Here, she kept the first account constant for those two weeks.

An alternative view of the conceptualization of "x" is promoted by a function-based algebra program where "x" is considered a changing quantity. In the context of this problem, the students whose instruction had been from this perspective seemed to recognize that the "x" stood for a dynamic and shifting amount of weeks that determined how much money each person had in their account: Muhammad had two scrolling tables in his head and tried to find the input that would have matching outputs (or as close to matching as we wanted). Nancy chose specific instances of possible inputs and computed to see if the outputs were equal, using this information to guide her next choice. Nathan was able to examine the graph of each function and, seeing that they increased at different rates, find the point where they intersected. In each of these cases, the students exhibited an understanding that for both people, the money in the account (the dependant variable) was changing at a predictable rate based on the change in time (the independent variable), and the question was therefore asking them to find the input where the two functions were equal.

The differences that these students presented in the mathematical approach they used to solve the banking problem has strengthened our belief that teaching algebra from a function-based approach is more likely to achieve the goals set forth in NCTM's Principles and Standards. It has also left us with additional questions, such as: How can we further evaluate this approach to teaching algebra? What, if any, are the drawbacks for teaching students to think of "x" as a variable quantity?

Difficulties with procedural knowledge. We observed differences in how students applied their procedural knowledge (computational and symbolic manipulation skills) in relation to the mathematical concept embedded within each of the three problems. Some students had a good connection between their procedural knowledge and what problems to apply it to, while others struggled to make connections between the process and what the question was asking for. Consider Jeff and Jamilla's responses as examples.

Jeff had no trouble solving the problem $2x + 10 = 100$, and gave an explanation that convinced us of his understanding. He knew while looking at the problem that the goal was "trying to find what x is" and was able to show why, when his process gave him $x = 45$, this answer made sense. When presented with $2x + 10 = 3x + 6$, he had a sense that this was quite different, because you have "got two variables." Taking some time to think he indicated that you would want "to put it into one." To do this, he began drawing arcs connecting the 2x and 3x, the 2x and 6, etc. Beneath the problem on his paper, he wrote, "$6x + 12x + 30x + 60$." The panel recognized he had tried to multiply the two linear functions. When we asked him what made him think to do that, he replied that it was what his teacher taught him to do, and because it was a "rule" that had a name he couldn't recall. From this point, he "combined the x's" writing "$48x + 60$" and after a long pause, he declared this was the "answer." He later explained that a learning strategy for him was to write mathematical rules down on note cards, and "if I read them every night, that's how I learned to do it." He said that he wished he had brought the note cards to the interviews.

Jamilla listened to the bank problem, and decided to subtract 220 from 500 by hand. She got a result of 480. She took that answer, divided it by the 30 that is being added per week to get the number of weeks asked for. She performed long division, and got the answer of 16. She put it all together and told us that it took 16 weeks to reach the $500 goal. Seeing that her reasoning was correct, but had a mistake in her calculation, we asked her how she could check so that she knew she was right. She smiled and, by hand, multiplied 16 by 30, getting the 480. With that, she confidently explained that she must be correct.

Jeff, when presented with $2x + 10 = 40$ and asked to solve it, was very good procedurally; i.e., he was able to subtract 10 from both sides accurately, keep track of the signs, isolate the variable, and check his answer. When presented with $2x + 10 = 3x + 6$, a procedurally similar problem, he had difficulty determining what he was supposed to do as a first step. Jeff was not aware that he was still just looking for a specific input where the two functions, $2x + 10$ and $3x + 6$, have equal outputs. Additionally, he tried to apply a simplification technique used to rewrite a product of linear binomials that is commonly taught as the FOIL (First, Outer, Inner, Last) method. Furthermore, even if it were appropriate to multiply the binomials in this situation, Jeff did not recognize that there was a problem with multiplying two non-horizontal linear functions together and getting another linear function. He was not aware of the difference between solving and changing the form of an expression, which illustrates a major gap in his conceptual understanding.

Jamilla felt it was sufficient to double-check one step in her two-step solution and when that was right, she assumed that the rest of her work was accurate. When she felt inclined to check her answer, she reverted to undoing her step to check, rather than trying to make sense of her answer. Jamilla had procedural knowledge of how to check her answers and undoubtedly had practiced it repeatedly. Her limited understanding of the checking procedure seemed to prevent her from extending it further to complete the process even with follow-up questions that provided her the opportunity to identify her error.

In both Jeff and Jamilla's approaches, it appears that an instructional over-reliance on procedures limited their ability to make sense of the problem and their own work. This mirrors what has been found in more comprehensive research that indicates that errors increase when procedures are not linked to a conceptual understanding (Hasselbring, Bottge, and Goin, 1992; Hiebert, 1986). In Jeff's case, some knowledge of procedures was there, but he was uncertain about when the procedures should be used. Indeed, it appeared that he was unsure what the procedures were supposed to do.

For Jeff, writing mathematical procedures on note cards and rotely practicing them multiple times was beneficial in remembering them, but was insufficient in helping him understand when the procedures could be appropriately applied. This seems to support NCTM's statement that "Students who memorize facts or procedures without understanding often are not sure when or how to use what they know, and such learning is often quite fragile" (Bransford, Brown, & Cocking, 1999, in NCTM *Principles and Standards*, p. 20). As the interview team, we were struck by the way that time spent in repetitive practice could actually work against the teachers' instructional goals for his or her students.

A common assumption, especially among those teaching students with learning disabilities, is that a lack of understanding is due simply to a longer processing time. And to some extent, Geary's (1994) research supports this assumption. Based on what we saw in the interviews, we have come to believe that the remedy, however, is not a diluted curriculum with more time to practice procedures that lack meaning for most students. Jeff and Jamilla had both been in remediated courses in their high school, yet, consistent with other research, this remediation and repetition had not helped them make sense of mathematics (Woodward & Montague, 2002).

These interactions raised a number of questions for us. What assumptions would we have made about Jamilla's understanding, based on her incorrect answer, if we hadn't taken the opportunity to talk to her about her thinking? How do teachers know when any student (let alone a classroom full) deeply understands the mathematics instead of having just memorized and repeated what the teacher wants to hear? What are the costs and benefits of teaching procedural knowledge without teaching students in which situations these procedures are appropriate?

Technology as a tool for learning. The interviews highlighted differences in the students' willingness to use a graphing calculator as a tool to help them think about the questions. One general observation that the team made during the student interviews was that several students were very tentative about using the calculator to solve problems, even though we made it readily available and were explicit that the graphing calculator was a tool that they could use. Others immediately used the provided TI-83 as a tool. For example, Martin graphed two functions on the TI-83, and was proficient using the zoom features and changing the display of one of the functions to distinguish it from the other.

The following excerpt illustrates how Muhammad's conceptual understanding of what it means to solve and his comfort level in using the graphing calculator provided him with the flexibility to choose a non-standard method to solve the second question in the banking problem:

> After using his table to find out that the accounts would be equal between week 5 and week 6, Muhammad said he could use the calculator to find exactly where they are equal. By hand, he thought to "subtract 140 or 220 from either side, maybe" but was unable to remember the process. After mentioning that interest would be really important to consider in this situation, he continued to solve the problem by putting $(45*x) + 140$ into y_1, and $(30*x) + 220$ into y_2. He then adjusted the increments between the inputs on the table and

asked if we wanted him to continue to change the increments to find a more accurate answer.

Muhammad was unable to remember how to symbolically solve the banking problem, though he did know what answer that process would provide. Instead, he turned to the calculator and used its calculating power to generate tables that he could use to find an answer to any given degree of accuracy. Muhammad's apparent comfort with using the graphing calculator as a tool allowed him to approach the problem in a way that the other students did not.

In contrast to Martin and Muhammad, it seemed obvious from the onset that other students' previous teachers had not viewed technology as a valid pedagogical tool in their classroom. For example, Jeremiah was so reluctant to pick up the calculator that he performed sixteen calculations by hand before using the calculator to check one of these answers (see vignette in the next section). For many, there were noticeable hesitations as they tried to gage from us if using the calculator was appropriate. One interviewer noted that a student "seemed to feel guilty for using the calculator."

In the *Principles and Standards for School Mathematics*, NCTM (2000) advocates for the use of technology, and for corresponding changes in teaching. Similarly, Woodward and Montague (2002) write:

> The range of computing tools, from handheld devices to powerful computers, will increasingly be used to perform the very computational functions that students have practiced on a daily basis by hand. These tools already allow today's students (and workers) to operate on data at a much higher level (e.g., analyze it for trends, perform statistical operations, solve complex problems). (p. 92)

Even given this, the role of technology, particularly calculators, is still a contentious topic for many educators. Using an "idiot box" to find a solution or quickly do a computation is often looked upon and communicated to students as a weakness rather than as resourcefulness, efficiency, or a tool to gain a larger vision of the problem at hand.

Rather than focus on the computational algorithms that are typically part of finding a solution, technology allows students to free up intellectual room to delve more deeply into the mathematics. In some ways, Muhammad illustrated a "higher level" of understanding by having access to several methods of finding a solution, one of which involved using technology to perform many operations simultaneously for him. In this way, he could quickly compare both functions and find where they were equal rather than solve the equation symbolically for the same outcome.

Multiple solution strategies. This category resulted from observing how students solved problems when a particular solution or strategy was not specified. Jeremiah gave us an interesting look into what can happen when we allow students to do what makes sense to them in solving mathematics problems:

> Jeremiah read the first part of the banking problem. He began by writing the goal of 500 on the top of his paper, and then subtracted the 220 to find the resulting 280 left to go. He then subtracted 30 from this amount, getting 250. Again, he subtracted 30, to get 220; and repeated this process of subtracting 30 until he had only $10 left. He then counted the thirties in his calculations, answering that it would take 9 weeks. When we asked if he thought this was right, he responded by saying that you could also divide 280 by 30, but that would be "hard to do." We reminded him that the calculator was at his disposal, and he then used it to check his result by dividing.

One perspective on Jeremiah's work would be to admonish him for "wasting time" by subtracting over and over, but he was more willing to do what he understood rather than do what would have been a quicker calculation. He was able to visualize the $30 payments for 9 weeks before needing to make only a partial payment. In addition, it seemed that he was more convinced by this first strategy, rather than the resulting 9.333333 that the division with the calculator gave him. The list of subtractions illustrated for him exactly the amount that was still needed after each week's deposit. Although it required more time, this approach was more compelling and reassuring to him than the result of a calculator's operation.

Nancy approached the problem with a different strategy:

> After hearing the first part of the banking problem, Nancy picked an answer of 5 weeks, and checked how much she would have had in her account after that amount of time. When asked, she indicated that she had chosen 5 randomly. After calculating that there would only be $370 in the account at this time, she decided that the answer would be more than 5, and tried 9 weeks. Checking that possible answer, she got $490, answering that the solution was "maybe 9½ weeks. Ten weeks would be over by $20."

Nancy simply used a guess-and-check method and, with minimal estimation, got an answer that made sense to her in the context of the situation. She was able to complete this process remarkably quickly, and exhibited a full understanding of what was happening in the situation. Later, when given problem #3, she again began with a random guess, adjusting as needed depending on the result. In this latter case, she relied on a very complete understanding of what it means to test for an input when one linear function is set equal to another, and then adjust to test a better input.

Again, we are left with more questions than answers. If students can get the correct answer and understand the situation enough to feel confident that their reasoning and processes are correct, do we *ever* need to dictate the solution strategy? If we view our role as facilitating the development of students' mathematical reasoning, and then discount a viable solution strategy, aren't we are undoing the very intent of our teaching methodology? Is the risk of disempowering the ideas of students ever worth the teacher's need to impose one certain, and perhaps more efficient, solution strategy? Do guess-and-check methods like Nancy's illustrate a lesser or higher degree of mathematical understanding than carefully applying a memorized procedure?

Students' experiences in learning mathematics. A final category that emerged from our observations involved the apparent experiences students have had in learning mathematics that have affected their confidence in their mathematical ability. In our conversations with the students, we gained some insights about their mathematics teachers' expectations for what mathematical solutions are acceptable and the viability of the use of technology. In this section, we consider a broader category that came out of our conversations about educators' beliefs about the mathematical capabilities of students with disabilities and the resulting learned coping strategies that we observed these students attempting to use. This is best thought of as an "implicit curricula" (Eisner, 1985) of high school mathematics classes, in which students formulate expectations for mathematics education that helps them stay afloat in a foreign landscape.

As an example of this, several students were unsure what to make of the panel when all we did was listen and ask follow-up questions, giving no indication of "right" or "wrong." Many teachers notice that at-risk students, and those with disabilities in particular, often learn to play off a teacher's subtle facial expressions and body language as a survival tactic when they lack confidence. In the interviews, it was not uncommon for students to say, "So, was I right?" Implicit in this question is a belief that the primary role of teacher is as evaluator, not as one interested in the true mathematical ideas and thoughts of a learner.

Something else that students learn through an unintended curriculum is what to expect from problems they encounter in the context of a mathematics class. Jason, when working on the stock problem, at one point said, "This probably is not the right formula I should be using because [my answer's] not exact." Part of the "implicit curriculum" that he has learned is that problems are carefully contrived to give neat and tidy answers. How will being proficient in this implicit curriculum help him translate his mathematical reasoning to the problems in the real world with realistic and often untidy solutions? De Corte, Greer, and Verschaffel (1996) advocate for a "less is more" approach where students, particularly those with disabilities, work completely through one or two realistic problems in a lesson rather than repeatedly practice one type of "highly situated" problem. Using realistic problems could allow the teacher to focus the classroom discussion on strategies employed, multiple solutions, and the nature of the problem at hand.

A second area that surfaced regarding student experiences is the message that they receive about their learning capabilities when they are repeatedly placed in remedial mathematics classes. This is particularly true for students with disabilities placed in segregated settings. It could be argued that students experiencing different expectations from various teachers is a natural part of growing up and, in fact, can help promote notions of pluralism and diversity. On a different day, we might say that adjusting their expectations for different classes and teachers allows students to develop some degree of flexibility and social awareness. However, when the expectations are habitually low, there is no silver lining. Students tracked in remedial courses or placed in special education classrooms receive the subtle, but powerful, message that they cannot learn the regular or advanced curriculum. Jason provides an example:

> After Jason sat down and introductions were made, he explained that he moved to the high school he graduated from before his sophomore year. There, he was put into "algebra modified 1 and 2" for his 10th grade year, and then put into "algebra modified 3 and 4" for his junior year math experience. After some clarification, it became clear that this was algebra slowed down over two years for students with special needs. He barely looked us in the face as he told us what classes he took. "I wish I didn't do it that way," he sheepishly replied. He went on to say that he was put into geometry his senior year, but dropped it. As if to confirm his destroyed confidence, he told us that he has asked the community college "for a lower class" that he could do. He asked us, "What's below this?" as we looked at the courses LCC offered.

Jason's confidence in his ability to learn mathematics was diminished through his experiences in high school. When verbalizing his ideas in the stock problem (problem #1) and the 2x+10 problem (problem #3), he showed little confidence that what he was doing was worthwhile or beneficial. Even though the team felt he was symbolically strong, demonstrated by his ability to solve problems like 2x+10=3x+6 without hesitation, he did not express any confidence in his ability to be successful, even in developmental community college mathematics courses. As it turned out, after his third semester Jason dropped out of college.

Jeff provides another example of how teachers' and even counselors' perceptions about a student's mathematical abilities are institutionalized and dictate a student's academic access unless challenged by a strong advocate.

> At the beginning of the interview, Jeff shared with us the story of how he got into the courses he took in his high school. Jeff wanted to take Algebra II as a senior, since he planned to be an electrician. However, he was told by both his teacher and counselor that he could not take this class, as he would never be able to pass it. When he persisted, his teacher made it clear to his parents and administrators that it was *not* recommended and therefore he could not elect this class. Finally, the school relented, but required that his parents sign a waiver relieving the school of any responsibility for his grade in this class. As Jeff told his story, his body language and voice inflection indicated a shift from anger to vindication when he told us that he earned an A in that class.

With Jeff, the message by the school is explicit: As an educational authority, we do not believe that you, as a student with a disability, have the capability to perform at the mathematical level of a regular sophomore. However, if your parents insist, they must sign a statement acknowledging the school's position and assuming full responsibility for the outcome. Fortunately Jeff, unlike Jason, had advocates and a support system to counteract these low expectations.

It appears that the school was more driven by the unwarranted judgment they had made about Jeff's ability than by looking at his future plans, making a decision about the mathematics he needed to reach his goal, and helping him succeed. Whether through years of subtle messages or an overt event like this one, educators are eroding students' beliefs in their own capabilities. According to NCTM (2000), "The vision of equity in mathematics education challenges a pervasive societal belief in North America that only some students are capable of learning mathematics" (p. 12). What do these kinds of messages from educators do to students' self efficacy about their academic abilities and their future goals? In the case of Jason, he quit pursuing his goal of a college degree and substituted a lesser goal. Jeff, however, was undeterred and used the resources of his parents to advocate for his needs. What about those students who do not have strong advocates in or out of school? What is the school's role in determining who has access to knowledge?

Indicated Pedagogical Strategies

The information gained from the interviews and our subsequent work led us to identify some key pedagogical strategies for teaching students mathematics, particularly those with disabilities. Our study revealed, to an extent we hadn't realized before, the depth of students' lack of conceptual development, the damaging effects of an over-reliance on procedures, the lack of confidence and autonomy in students' problem solving, and the debilitating effects of school policies and teacher remarks on students' esteem. A heightened awareness of such issues now motivates us more than ever to change our teaching in ways that mitigate these problems. Additionally, the interview process has also highlighted some ideas about the questioning process and its potential pedagogical benefits for the classroom. In this section, we make a case for pedagogy that creates a classroom environment where students recognize themselves as the authority for thinking, talking about, and evaluating each others' answers, and that follows up on a student's comment or partial answer in the hopes of probing for the thinking or understanding that the student or class possesses.

In the months following the data collection, we had dynamic conversations among ourselves and with other professionals about mathematical pedagogy, matters of content authority in the classroom, and concerns about the mathematics education of students with disabilities (Huhn, Huhn, & Lamb, n.d.; Lamb, Jackson, Petry, Huhn, & Huhn, 2002). Specifically, the student interviews powerfully illustrated how follow-up questions can be used as a means of assessing student understanding and point to their potential for day-to-day classroom use. We saw evidence that telling does not work—students develop conceptual understanding by constructing mathematical connections for themselves. Through daily dialogue about the mathematical reasoning students are applying as they solve problems, we conjecture that students will develop a deeper understanding of mathematics that can be applied and recalled after the unit test. When students had misunderstandings, we saw how probing, provocative questions could assist them in identifying where their thinking may be off-track and in restructuring their reasoning process. We have come to believe that, as teachers, this means we need to respond to mathematical questions from students with questions all the time. We need to make a consistent effort to shift the mathematical authority to the students. If we relinquish that control sometimes, and yet try to wield it at other times, students will no longer engage in deep thinking activities. This means we need to let students be wrong for periods of time, even though that might feel uncomfortable. And we need to practice patience so students can be allowed to forge their own paths towards mathematical understanding.

We recognize that advocating for a classroom that looks like this requires a shift in the perception of who maintains mathematical authority in the classroom. It changes the traditional paradigm of the teacher being viewed as a source of mathematical knowledge, to a facilitator who guides the development of mathematical power and authority within the students themselves. It shifts the role of mathematical authority to the students, relying on them to decide what is "right" and which strategies work. We were well aware as we listened to students of the difficulty in watching them struggle. We often talked about how it was against our instincts as teachers to not step in and show them how to complete the problem. However, our teaching experience and our observations of students' thinking in these interviews has convinced us that if a teacher interrupts the mathematical struggle necessary for student learning and provides an answer, then it is almost impossible for students to resume the fight to learn to think mathematically. As a result, the opportunity for higher levels of mathematical understanding is compromised.

In order to use the clinical interview model in the classroom, teachers have to be very savvy and knowledgeable of their subject matter. It requires a strong mathematical background to be able to recognize the possible implications of students' conjectures and generalizations of solution strategies, as well as to handle their misconceptions. It may seem quite ironic that in a model of mathematics education in which the students do a great deal of the talking and thinking the teacher would need such depth of subject matter knowledge. However, this approach requires that teachers be very strong and confident in their own mathematical reasoning in order to facilitate the development of this reasoning in their students. They must be prepared to follow up on any relevant ideas or questions that a class has. The teacher needs to be very clear about the goals for the lesson and its place in the curriculum in order to choose appropriate follow-up questions. Perhaps most importantly, this takes a complete shift in paradigm: teaching is not the imparting of a third person's knowledge, nor is it the sharing of knowledge gathered in textbooks. Instead, it is facilitating the development of the students' mathematical understanding, which requires an openness of mind to allow the students to express their ideas and the skill to follow-up on them in productive ways.

Finally, in the interviews we noted several instances of the negative effect that pull-out special education classrooms and tracking have had on students with disabilities. Research has long documented the inequalities of education in pull-out situations, including lowered expectations, lower standards, a diluted curriculum, and teachers who are themselves limited in subject matter knowledge (Cunningham, 1998; Fuchs, Fuchs, & Hamlet, 1989; Gurganus, Janas, & Schmitt, 1995; Pugach & Warger, 1996; Wheelock, 1992). We witnessed, even among students who were choosing to follow careers in mathematics, science, technology or engineering, a lack of mathematical confidence that diminished their self-esteem.

Hearing the various stories at the onset of the interviews raised our awareness that even in our surrounding communities, access and opportunity are denied to students with disabilities on a regular basis. According to Goodlad (1990), the school is the only institution given the responsibility to provide youth with the knowledge and experiences in all subject matters. Therefore, teachers must be attentive that no belief or practice limits any student's access to knowledge. The moral purpose of school is to teach all students, and teachers need to strongly advocate for all students' rights to have access to mathematical knowledge. NCTM says very clearly, "structures that exclude certain groups of students from a challenging, comprehensive mathematics program should be dismantled" (p. 369). The problem that remains, then, is how to overcome the persistent beliefs and practices that disenfranchise students by limiting their access to mathematics.

ACKNOWLEDGMENTS

This work was funded by National Science Foundation Programs for Persons with Disabilities Grant #HRD9906043.

The authors wish to acknowledge the students who participated in the interviews and our mathematical colleagues, Nan Jackson and William Petry, from Lansing Community College. In addition, we would like to thank Dr. Daniel Chazan and Sandy Callis for professional development opportunities in mathematics education.

REFERENCES

Bransford, J., Brown, A., & Cocking, R. (Eds.). (1999). *How people learn: Brain, mind, experience, and school.* Washington, DC: National Academy Press.

Chazan, D. (1993). F(x) = G(x)?: An approach to modeling with algebra. *For the Learning of Mathematics, 13,* 22-26.

Chazan, D. (1999). On teachers' mathematical knowledge and student exploration: A personal story about teaching a technologically supported approach to school algebra. *International Journal of Computers for Mathematical Learning, 4,* 121-149.

Cunningham C. (1998, March). *EASI Street to science and math for K–12 students.* Paper presented at a Persons with Disabilities Conference, Los Angeles. Retrieved July 2, 2002, from http://www.dinf/org/doc/English/

De Corte, E., Greer, B., & Verschaffel, L. (1996). Mathematics teaching and learning. In D. C. Berliner & R. C. Calfee (Eds.), *Handbook of Educational Psychology* (pp. 491-549). New York: Macmillan.

Eisner, E. W. (1985). The three curricula that all schools teach." In *The educational imagination: On the design and evaluation of school programs* (pp. 87-107). New York: Macmillan.

Fuchs, L. S., Fuchs, D., & Hamlet, C. L. (1989). Curriculum based measurement: A methodology for evaluating and improving students programs. *Diagnostique, 14,* 3-13.

Geary, D. (1994). *Children's mathematical development*. Washington, DC: American Psychological Association.

Goodlad, J. (1990). *Teachers for our nation's schools*. San Francisco, CA: Jossey-Bass.

Gurganus, S., Janas, M., & Schmitt, L. (1995, Summer). Science instruction: What special education teachers need to know and what roles they need to play. *Teaching Exceptional Children*, pp. 7-9.

Hasselbring, T., Bottge B., & Goin L. (1992). *Linking procedural and conceptual knowledge to improve mathematical problem solving in at-risk students*. Paper presented at annual meeting of the Council for Exceptional Children, Baltimore.

Heid, K. (1996). A technology-intensive functional approach to the emergence of algebraic thinking. In N. Bednarz, C. Kieran, & L. Lee (Eds.), *Approaches to algebra: Perspectives for research and teaching* (pp. 239-255). Dordrecht: Kluwer.

Heid, M. K., Choate, J., Sheets, C., & Zbiek, R. M. (1995). *Algebra in a technological world*. Reston, VA: NCTM.

Hiebert, J. (1986). *Conceptual and procedural knowledge: The case of mathematics*. Hillsdale, NJ: Erlbaum.

Hubbard, R., & Power, B. (1999). *Living the questions: A guide for teacher-researchers*. Portland, ME: Stenhouse.

Huhn, K., Huhn C., & Lamb, M. (n.d.). A retrospective view on the importance of elementary experiences in mathematics for the transition of students with disabilities into the workforce of the twenty-first century. Unpublished manuscript.

Lamb, M., Jackson, N., Petry, W., Huhn, C., & Huhn, K. (2002, November). *NSF Bridges Project: Transitioning youth with disabilities into mathematical careers*. Paper presented at a regional conference for National Council of Teachers of Mathematics, Boston.

Lipsky, D. L., & Gartner, A. (1997). *Inclusion and school reform: Transforming America's classrooms*. Baltimore: Brookes.

National Council of Teachers of Mathematics. (1989). *Curriculum and evaluation standards for school mathematics*. Reston, VA: Author.

National Council of Teachers of Mathematics. (1991). *Professional standards for teaching mathematics*. Reston, VA: Author.

National Council of Teachers of Mathematics. (2000). *Principles and standards for school mathematics*. Reston, VA: Author.

Oakes J. (1985). *Keeping track: How schools structure inequality*. New Haven and London: Yale University Press.

Pugach, M., & Warger, C. (Eds.). *Curriculum trends, special education, and reform refocusing the conversation*. New York: Teachers College Press.

Wheelock, A. (1992). *Crossing the tracks: How "untracking" can save America's schools*. New York: New Press.

Woodward, J., & Montague, M. (2002). Meeting the challenge of mathematics reform for students with LD. *The Journal of Special Education, 36,* 89-101.

Yerushalmy, M., & Gilead, S. (1997). Solving equations in a technological environment: Seeing and manipulating. *Mathematics Teacher, 90,* 156-163.

CHAPTER 7

NAVIGATING THE LEARNING CURVE

Learning to Teach Mathematics Through Lesson Study

John Carter
Adlai E. Stevenson High School

Robert Gammelgaard
Adlai E. Stevenson High School

Michelle Pope
Adlai E. Stevenson High School

The difficulty that experienced teachers often face when making fundamental changes in their teaching practices highlights the need for novice teachers to establish productive patterns early in their career. It is well established that, "left on their own, beginning teachers will focus on whatever works.... [A]ttitudes tend to become more negative and behavior becomes more inflexible" (Brock & Grady, 1997, p. 10). The literature

Teachers Engaged in Research
Inquiry Into Mathematics Classrooms, Grades 9–12, pages 119–134

makes it clear that professional development of new teachers must begin as soon as possible in their careers (e.g., Brock & Grady, 1997; Lambert, Collay, Dietz, Kent, & Richert, 1996). In particular, novice teachers need induction experiences that are subject-specific, rather than generic first-year teacher strategies, so that they can explore pedagogical strategies that will work in their specific contexts (Darling-Hammond, Berry, Haselkorn, & Fideler, 1999).

Schools that are committed to implementing the NCTM *Standards* (1989, 1991, 1995, 2000) face additional challenges when inducting new teachers. Being asked to teach mathematics in a manner that most have not directly experienced as learners often creates a tension between novice teachers' perceptions of and the school's expectations for how mathematics should be taught. One way to address this tension is to carefully craft induction programs that offer novice teachers the opportunity to reflect on teaching in light of the needed reforms. Doing so requires a conscious effort by experienced department members to teach the tenets of teacher collaboration, collective inquiry, and experimentation, and to develop habits of continuous improvement (DuFour & Eaker, 1998).

Very few schools, however, have developed teacher induction programs that do this. One possibility that our school has explored is involving new teachers in an induction program that uses Lesson Study as a means of professional development. We wanted to explore the potential of Lesson Study to help novice teachers learn to work together as they make sense of the requirements of the mathematics department with regard to lesson design and implementation. In this chapter we describe how we used the Lesson Study component of a larger induction program to focus novice teachers' attention on meeting the goals of the mathematics department early in their teaching career.

LESSON STUDY

Lesson study is a professional development experience in which a group of teachers works collaboratively to: formulate a goal for student learning; plan and teach a lesson focused on the goal; devise and implement a strategy for assessing progress toward the goal; and revise and re-teach the lesson. As a part of the lesson study cycle, one member of the group teaches the lesson while all other participants observe and collect data. After the teachers meet to discuss the lesson and make revisions, a member of the group teaches the revised lesson while all members observe and collect data for subsequent collaborative analysis. This process is summarized in Figure 7.1.

Steps of the Lesson Study Process

1. Collaboratively set lesson goal
2. Research lesson topic and methods
3. Plan lesson
4. Teach and observe lesson
5. Evaluate lesson
6. Revise and re-teach lesson

Figure 7.1.   Steps of the Lesson Study Process.

Lesson Study provides the context for needed reflection, dialogue, and an interactive professional culture. This type of professional development affords teachers the opportunity "to fashion new knowledge and beliefs about content, pedagogy, and learners" (Darling-Hammond & McLaughlin, 1995, p. 597). Through intense collaborative sessions the participants study every component of the focus lesson. The professional exchanges between group members help each participant to articulate ideas, opinions, and perceptions. The group's data collection and debriefing sessions allow the teachers to compare what transpired in the classroom with their anticipated sequence of events. The process of Lesson Study engages teachers in research through reading other's research, integrating what was learned into their lesson, and investigating the results.

BACKGROUND

The Context

Stevenson High School is a large comprehensive public high school in a suburban setting. The mathematics department consists of 44 mathematics teachers. More than 95% of the 4,500 students continue their formal education after high school. Stevenson High School is guided by the tenets of a Professional Learning Community (DuFour & Eaker, 1998). The school culture is guided by a commitment to the district vision for education that includes teacher collaboration, a results-orientation, and a focus on continuous improvement. The curriculum at Stevenson is a set of departmental learning expectations for each course coupled with a departmental vision of instruction. The teachers of each course come to a consensus on how that will be achieved on a course-by-course basis and work together to implement their plan. Within the mathematics department, the professional development of teachers is focused on the vision of mathematics education as outlined in the *Principles and Standards for School Mathematics* (NCTM, 2000). The entire department meets five times per year to engage in discussions regarding current research in mathematics education and its applicability to our department.

In addition to the departmental professional development, since the 2001–2002 school year mathematics teachers in their first and second year of teaching at Stevenson High School have met monthly for half of a school day to read and discuss current publications in mathematics education. As a part of each of two required iterations of the year-long induction program, the group of teachers engaged in a Lesson Study, as described above, with the goal of implementing the reforms called for in their readings. That is, each teacher completed two cycles of the Lesson Study process outlined in Figure 1.

Hiebert et al.'s (1997) work on classrooms that promote understanding (see Table 1) provided a framework for our professional development program. Reading excerpts from this text before meeting provided focus for our discussions. We concentrated on making sense of the dimensions and core features of classrooms that promote understanding in relation to our own situations and consistently referenced the summary in Table 7.1 throughout our Lesson Study process.

The Participants

During the 2001–2002 school year, two of the authors, Robert and Michelle, participated in our department's first Lesson Study with five other mathematics colleagues. The participants were all first-year teachers at Stevenson High School with teaching experience ranging from 0 to 5 years. Three of the participants had graduated from teacher preparation programs specifically focused on the NCTM's Standards. The other participants graduated from programs in which the NCTM Standards were presented, but were not the identified focus. The Lesson Study was facilitated by the third author, John, the director of mathematics and an experienced member of the department. The Lesson Study served as the final phase of the new teacher induction program for the mathematics department. The remainder of this paper is the story of our Lesson Study experience told through the eyes of Robert and Michelle, with input from John.

Table 7.1. Summary of Dimensions and Core Features of
Classrooms that Promote Understanding

Dimensions	Core Features
Nature of Classroom Tasks	Make mathematics problematic Connect with where students are Leave behind something of mathematical value
Role of the Teacher	Select tasks with goals in mind Share essential information Establish classroom culture
Social Culture of the Classroom	Ideas and methods are valued Students choose and share their methods Mistakes are learning sites for everyone Correctness resides in mathematical argument
Mathematical Tools as Learning Supports	Meaning for tools must be constructed by each user Used with purpose – to solve problems Used for recording, communicating, and thinking
Equity and Accessibility	Tasks are accessible to all students Every students is heard Every student contributes

Reprinted by permission from *Making Sense: Teaching and Learning Mathematics with Understanding* by J. Hiebert et al., copyright © 1997 by the University of Wisconsin Foundation. Published by Heinemann, a division of Reed Elsevier, Inc., Portsmouth, NH. All rights reserved.

PARTICIPATING IN THE PROCESS

As a part of the departmental induction program, we had five monthly sessions where we were provided substitutes to allow us to meet for half of our work day to work on our Lesson Study. These monthly sessions augmented the regularly-scheduled departmental professional development meetings and allowed us, the 7 teachers new to the department, to focus on making sense of the departmental expectations and culture. In the following sections we first address preparing for the Lesson Study and then describe our participation in each part of the six-step Lesson Study process in more detail.

Preparing for the Lesson Study

As teachers who were novices to the school's collaborative environment, we needed to learn more about how to function well as a collaborative group of professionals. An important part of this was discussing our expectations of each other. For example, we felt it was important for everyone to provide input and for everyone to listen when input was being given. Most of our group norms emerged as we learned to work with one another. In subsequent Lesson Study Cycles, group norms have been discussed and recorded from the start. These norms generally state the group's expectation regarding honoring commitments to the group, active listening, equitable participation, acknowledging experiences of peers, and remaining supportive of each other during this time of growth.

Another critical component was time spent discussing the preparatory readings at length. We challenged each other's understanding of the research by asking for clarification or examples of how the speaker envisioned these ideas in his or her classroom. These discussions gave us insight into each other's past experiences and the way in which we conducted our various classes. They allowed our group to become familiar with each other's strengths, weaknesses, beliefs, and personalities, and to appreciate the unique characteristics each person brought to the group. The time taken to establish our norms for participation and to share common experiences enabled us to build a level of trust that served us well during the Lesson Study process.

Setting the Goal

Using what we had read about the Lesson Study process, we began with setting a goal for our work together. At the group's first session, we decided to focus our efforts on promoting learning for understanding as described in the preparatory readings. With Heibert et al.'s (1997) dimensions and core features of classrooms that promote understanding (see Table 7.1) as our framework, we were able to discuss and analyze our current beliefs and practices. A common theme was changes that would have to be made in our teaching behaviors in order to reach our goal of teaching for understanding. For example, we discussed the need to provide students with more wait time as we shifted to a task-based classroom. In another conversation, members of the group expressed concern that teaching via tasks would limit the amount of time they could dedicate to working out examples for students. This was a concern because of the requirement that teachers cover the entire curriculum and the belief that providing worked examples was an efficient way to do that.

As a part of our early discussions we struggled with what learning for understanding meant and entailed. Initially, when our students were able to reproduce something taught in class, we thought that meant they understood the concept. Reading research about best practices challenged our beliefs about student learning. We took from the readings that learning for understanding involved making connections to things already known. The problem we encountered was how to help students make those connections.

Our goal was to develop a lesson that incorporated the type of student engagement, reflection, articulation, and communication emphasized in the readings we had completed on teaching for understanding. We wanted to choose a topic that would allow us to make progress toward that goal. Given our varied experiences, some members felt certain topics would be too difficult to tackle in our first attempt. Other members felt that some topics were too easy and our group would not be appropriately challenged to meet our goal. Our selection was also strongly influenced by the topics scheduled to be taught in the spring. In the end, we decided that writing a lesson to introduce exponential growth and decay would meet all our criteria.

Researching and Creating the Lesson

The second meeting focused on refining the content goals of the lesson and identifying what we felt was the important mathematical content that the students should understand. We wanted to create a lesson that would contribute to meeting our instructional goals, allow students to make connections to prior knowledge, and be mathematically problematic and interesting to students. We realized that trying to incorporate these ideas into one lesson could be daunting. For many of us, this was our first experience applying the research-based framework that we had discussed. As the research indicated, we needed to design tasks that would pique student interest and "leave behind something of mathematical value" (Hiebert et al., 1997, p. 12). As we worked to create the tasks that we would use in our lesson, we were mindful of the fact that part of the framework from which we were working required that a social culture be created in the classroom that would allow students to share ideas and methods and learn from each other.

Based upon our goal, we identified a need to create tasks that would help students build an understanding of exponential functions. We also wanted the mathematics in the lesson to be problematic and engaging. In order to create these tasks, we realized that we first had to identify what was important about exponential functions and when in the unit the lesson would be taught. We decided to create a task that would allow students to consider the differences between exponential growth and linear growth by building on their prior knowledge of linear functions. We wanted to create a situation that allowed students to build the growth pattern by repeatedly multiplying by the growth factor. At the time, we anticipated students would generalize this process and recognize the need for exponents. As we designed the tasks and lesson, we continually referred back to the readings, particularly concentrating on the five dimensions outlined by Hiebert et al. (1997) (see Table 1), and revised the tasks accordingly. For example, when considering the role of the teacher and the social culture of the classroom, we decided to leave the correctness of the opening activity to be discovered in the mathematics of a later task. The readings provided a filter for how we came to analyze our work. We regularly revisited the dimensions and core features of a classroom that promotes understanding and, as a group, discussed our own progress in each area.

After several hours of discussing, brainstorming, and completing potential tasks in our notes and on the board, we started to assemble a lesson plan. We chose to begin the lesson by asking our groups of students to determine the amount of money they would have after 365 days if they accepted a contract to work for $0.05 on the first day and, after each day worked, they exchanged their previous earnings for 106% of what they had earned since the beginning of the contract. We agreed on this question because we felt it would allow us to elicit student thinking about exponential growth.

Much time was spent framing the questions we wanted to ask and anticipating student responses. As we became more aware of each other's thinking, we found ourselves generating multiple hypotheses as to what the students would do in response to the task. For example, some members of the group hypothesized that students would calculate 6% of $0.05, multiply it by 364 and add it to the original. Others felt that students would multiply $0.05 by 1.06, 364 times. Still others predicted students would multiply by 6% and add the result to $0.05 and repeat the process as often as needed. Due to the comfort level of the group and the focus on student thinking, no one expressed fear about contributing what may have been incorrect mathematical approaches and we were able to work through our own incomplete understandings. This type of discussion about potential strategies provided us with the opportunity to think about how we would respond to the groups of students who produced this kind of work. We agreed that one of our responses would be to ask students to explain their reasoning.

During the next four meetings, we continued to develop the lesson. As the participants shared their ideas, these ideas would often trigger a new idea for someone else. By the conclusion of the second session, we had written a warm-up activity and had begun to create a framework for the task that would serve as the foundation for the lesson. We knew we wanted a task-based lesson and had to agree on how we would implement this approach that was new for all of us as novice teachers.

In creating the lesson, it became apparent that we had different views of how to implement the research-based practices. As facilitator, John encouraged us to work through these issues and come to consensus about how to proceed. When these differences were encountered, we listened to each other's ideas and determined what we thought would best help us to implement Hiebert's framework. We discussed the extent to which we would provide information to students and how we would respond to their questions and work. We discussed various proposals about how students would respond to the questions that were posed. Ideas and beliefs regarding calculator usage and how to implement it in our lesson helped to further shape our plans. Concerns arose regarding answering student's questions. We found ourselves challenged by the analysis of the locus of control and sources of mathematical authority. We talked about our own classroom habits and referred back to Hiebert et al. (1997) as a way to examine what we were currently doing in our own classrooms versus what we wanted to do in the lesson study. These conversations caused us to reflect upon and often begin to modify our own teaching and practices.

In order to better understand possible student thinking, we concentrated on predicting student responses as a basis for developing our task questions. How would they write their expressions? Would they use words to describe their calculator methods? Would they write numeric expressions? Would someone see the pattern right away and show their group members before they had a chance to discover it for themselves? We tried to think of every possible student response and then decide how to structure the questions based on these responses. For example, we decided to ask a question that required students to *think* about and write out what they would enter into their calculator to solve a problem instead of actually using their calculator. We developed a sequence of instructions and questions that followed this idea of thinking and writing before typing into a calculator in the hope that students would begin to see a pattern within their work.

TEACHING THE LESSON

By the end of our fifth meeting, we felt that the lesson aligned with the dimensions and core features of mathematics classrooms that promote understanding. The next step was to teach the lesson and gather data about how well our goals were actually achieved. One member of our team taught the lesson while the rest of us observed and collected data. Instead of a typical emphasis on the teacher, the observational data centered on the ways the students thought about and responded to the task and questions. Students were seated in groups of four, and one observer was assigned to each group. Observers took notes on student interactions, calculator usage, and what students wrote on their paper. Early in the process, the group had agreed on who would teach the lesson during this phase of the process. Two of the group members agreed to be the teachers for our collaboratively planned lesson. We found it interesting that *who* taught the lesson was of less importance to us than *how* the lesson was taught. We all knew that the lesson was *ours* and how one of us approached it was a direct result of what *we* had developed.

In order to collect detailed data about student thinking, each observer recorded the sequence of student thinking and responses to task questions and prompts so that we could compare students' responses with our hypotheses of how they might think about the task. We were interested in determining if we had predicted the solutions and thinking that students produced. We also wanted to know how student thinking progressed and when it became formal enough to move to the use of variables.

Afterwards, we met for one hour to discuss and agree upon any modifications. We discussed the data collected by each observer and compared it to our anticipated results. One change we felt necessary was adding additional time for students to think and write individually before allowing group discussion. We restructured the teacher portion of the lesson to allow for this additional time. We also clarified our own expectations with respect to calculator usage. We then watched the modified lesson being taught again by a different member of the group. We reconvened once more to evaluate the second lesson and discuss our overall progress.

An important part of the process for us was to compare our predicted paths for student learning with what actually transpired based on the data collected around the room. In the debriefing sessions we shared those observations and discussed how they compared. In this case, the observers noted that we had predicted almost all methods used by the students. We had already discussed how to handle certain responses, and how to help students understand one another's ideas. However, during the first observation we were surprised by a few of the students who were able to quickly write the correct expression, but were unable to explain their reasoning.

As we reflected on the first observation, we spent time discussing that students are sometimes able to write the correct answer, but do not always have a complete understanding of the concept. We revisited the idea of encouraging students to explain their work, *even* when their answer appears correct. Our ideas were then implemented into the lesson for the second observation.

LESSONS LEARNED

Role of the Teacher

As we worked and grew together, we all came to develop and articulate strong opinions about the role of the teacher. Finding the balance between giving students necessary information and guidance and demonstrating a predetermined process and ordaining it as correct proved to be a challenging component of our instructional goal. Our role in the classroom shifted towards empowering students to develop their own way of thinking, while leading them to a deeper understanding of important mathematical concepts. We agreed that we needed to stop telling students directly if they were correct. Instead, we needed to refer students back to mathematical argument as a means for determining correctness. We hoped to facilitate discovery and discussion, and, as a result, create a classroom culture where this would be the norm—one in which students regularly displayed and explained their own answers. We anticipated the students' methods and possible mistakes and wanted them all to be valued and appreciated by their peers. It was a learning experience for most of us to discuss how we could make this happen.

The Instructional Environment

Initially, we did not understand how much the instructional environment impacted equity in the classroom. We noticed the importance of having a task that was accessible to all students and of preventing the thinking of one student from trampling on the thoughts of another. By assuming the student's role and working through the mathematics, we came to recognize the need for uninterrupted time for students to think about the task. Reflecting upon our own use of wait time, together we devised a slightly different approach that is still visible in many of our classrooms. For example, in our discussion we all expressed concern about students calling out responses to questions before others had a real opportunity to think deeply about the questions we had designed, a practice that had the potential to create inequities in our classrooms. Thus, we decided that whenever we would ask a question that we felt merited a reflective response, we would direct students to think about the question for a period of 30 seconds in absolute silence. After the established "think time" had expired, we would direct students to share their thinking with the other students in their group. After the think time and social processing, we would then ask for responses from the groups of students. For us, this sent a message that thinking was valued and that we recognized the need to process information in a small group before sharing it with the entire class.

Our thoughtful discussions in the Lesson Study started to impact our classrooms in additional ways throughout the process. We thought about the framework established by Hiebert when developing tasks for our daily lessons. We created environments that fostered the sharing and exploring of students' ideas and methods. All participants reported these types of changes in their own teaching behavior through the reflective journals that they wrote. John also noticed these changes as he observed lessons throughout the year in his role as the director of mathematics. In comparison with novice teachers in the past, novice teachers who had participated in Lesson Study designed more student-centered lessons and were better able to analyze their own lessons with respect to fostering student discourse.

Reflecting on Practice

Reflecting on the lessons in the context of our Lesson Study group helped us all learn to critique each other and ourselves in light of our stated goals. We learned we had to challenge each other to support any and all observational claims. The relationships we had developed during the year helped us all to feel more comfortable critiquing our own performance. In retrospect, we attribute this largely to the norms we established at the beginning of the experience. We had agreed to be active listeners, withhold judgment on new ideas, respect each other's strengths and needs, and not allow any one person to dominate the discussion. While the actual group norms are not surprising, our stake in their creation and articulation allowed us to hold each other to them.

A facet of Lesson Study that was difficult to appreciate while immersed in it, but became clear through our reflection, is the fact that the collaborative process mirrors our own expectations for student learning. As we worked to design a lesson in which students were actively engaged, sharing ideas, listening carefully, making conjectures, and collecting data to inform decisions, we were doing exactly that ourselves—as a group of teachers we were engaged in our own professional learning.

Benefits Outweigh Difficulties

Integrating Lesson Study into the teacher induction process is intense. Our district has now repeated the Lesson Study with several groups of novice teachers. Throughout the process each year, participants report a variety of feelings. Journal entries over the years have included statements of fear, excitement, anger, elation, weariness, frustration, and success. The professional interaction and high level of focus can make for a long day if it occurs after school. For that reason, we chose to meet for half school days throughout the year. One of the difficulties with this approach was that it required teachers to miss time with their students. While this is always difficult, we have seen evidence that the resulting professional growth outweighs the time missed.

Although a common misunderstanding about Lesson Study is that the time invested is limited to improving one lesson, we have seen evidence that the effects of the process reach far beyond a single lesson and potentially far into the fabric of a novice teacher's career. Based on John's observations and those of the other school administrator who evaluates the new mathematics teachers, immersion in the Lesson Study process has helped each teacher make progress that is usually not witnessed until later in the typical teacher's career. Very important to the goals of our department, these teachers demonstrated more confidence and skill in implementing student-centered lessons focused on student discourse.

CONCLUSION

The purpose of the Lesson Study component of our department's induction program is to help novice teachers learn to work together as they make sense of the requirements of the mathematics department with regard to lesson design and implementation. Since experienced teachers often have difficulty making fundamental changes in teaching practices, it is important that novice teachers do not establish unproductive patterns early in their career that may later be difficult to change. Intervening early to help novice teachers develop a shared understanding of the departmental expectations can help to sharpen the focus of their teaching and promote the development of healthy instructional patterns. Providing novice teachers the opportunity to construct an early understanding of lesson characteristics that promote student understanding (Carpenter & Lehrer, 1999; Hiebert et al., 1997) through collaboration with their peers holds promise for accelerating their own learning along an otherwise steep learning curve.

Bringing a group of novice mathematics teachers together to collaboratively plan a lesson proved to have multiple benefits that we feel are generalizable. The group became a team and learned to work together toward their common goal. The camaraderie that developed helped each novice teacher develop the confidence needed to implement new pedagogical techniques in his or her classroom. In addition, the collaborative atmosphere helped counteract the feelings of isolation that often contribute to teachers leaving the field within the first three years of teaching (Gordon & Maxey, 2000). The detailed lesson planning process helped all members of the group to deepen their content knowledge as they faced common misconceptions or difficulties together. It also resulted in the novice teachers learning more about the department's curriculum and culture. The discussions that ensued about students' prior knowledge helped to broaden the perspective of all the participants with respect to the work of the department.

Based on our experience, we believe that this kind of induction program has the potential to challenge and modify beliefs that interfere with the goal of teaching mathematics for understanding. We encourage others to use Lesson Study as a way to engage teachers in research-based learning communities at their school. The collaborative nature of Lesson Study and focus on student thinking can accelerate the development of novice teachers by providing them with the learning experiences they need to construct an understanding of reform-based mathematics teaching and to implement it in their classrooms.

REFERENCES

Brock, B. L., & Grady, M. L. (1997). *From first-year to first-rate: Principals guiding beginning teachers.* Thousand Oaks, CA: Corwin Press.

Carpenter, T. P., & Lehrer, R. Teaching and learning mathematics with understanding. In E. Fennema & T. A. Romberg (Eds.), *Mathematics classrooms that promote understanding* (pp. 19–32). Mahwah, NJ: Lawrence Erlbaum Associates.

Darling-Hammond, L., Berry, B., Haselkorn, D., & Fideler, E. (1999). Teacher recruitment, selection, and induction: Policy influences on the supply and quality of teachers. In L. Darling-Hammond & G. Sykes (Eds.), *Teaching as the learning profession: Handbook of policy and practice* (pp. 183–232). San Francisco: Jossey-Bass.

Darling-Hammond, L., & McLaughlin, M. W. (1995). Policies that support professional development in an era of reform. *Phi Delta Kappan, 76*(8), 597–604.

DuFour, R., & Eaker, R. (1998). *Professional learning communities at work: Best practices for enhancing student achievement.* Bloomington, IN: National Educational Service.

Gordon, S. P., & Maxey, S. (2000). *How to help beginning teachers succeed.* Alexandria, VA: Association for Supervision and Curriculum Development.

Hiebert, J., Carpenter, T. P., Fennema, E., Fuson, K. C., Wearne, D., Murray, H., Olivier, A., & Human, P. (1997). *Making sense: Teaching and learning mathematics with understanding.* Portsmouth, NH: Heinemann.

Lambert, L., Collay, M., Dietz, M. E., Kent, K., & Richert, A. E. (1996). *Who will save our schools? Teachers as constructivist leaders.* Thousand Oaks, CA: Corwin Press.

National Council of Teachers of Mathematics. (1989). *Curriculum and evaluation standards for school mathematics.* Reston, VA: Author.

National Council of Teachers of Mathematics. (1991). *Professional standards for school mathematics.* Reston, VA: Author.

National Council of Teachers of Mathematics. (1995). *Assessment standards for school mathematics.* Reston, VA: Author.

National Council of Teachers of Mathematics. (2000). *Principles and standards for school mathematics.* Reston, VA: Author.

CHAPTER 8

LEARNING FROM ELEMENTARY SCHOOL MATHEMATICS RESEARCH:

Changes in the Beliefs and Practices of Secondary School Teachers

Scott Hendrickson
Brigham Young University[1]

Sharon Christensen
Mountain Ridge Junior High School

with contributing research by

Vicki Lyons
Lone Peak High School

Adrianne Olson
Lone Peak High School

For the past several years we have played a dual role in the field of mathematics education. Each morning Scott teaches algebra, precalculus, and Advanced Placement calculus classes at a local high school. At the same

Teachers Engaged in Research
Inquiry Into Mathematics Classrooms, Grades 9–12, pages 135–152
Copyright © 2006 by Information Age Publishing
All rights of reproduction in any form reserved.

time, Sharon teaches prealgebra and algebra in a feeder junior high. In the afternoons, we work as mathematics specialists in our school district. Our dual assignments give us two perspectives on mathematics education reform: one from the position of training and supporting teachers as they make the transition from traditional ways of teaching mathematics toward developing new pedagogical strategies, and one from the position of teachers in the trenches trying to make sense of the new pedagogy for ourselves.

As our district prepared to adopt mathematics curricula based on the National Council of Teachers of Mathematics *Standards* (1989, 1991, 1995, 2000), we organized a K–12 Mathematics Alignment Committee to study the current research on how children learn mathematics. Participants—consisting of four elementary, four junior high, and four high school teachers, including the district elementary and secondary school mathematics coordinators—were selected to serve on the committee based on their initial efforts to change the nature of mathematics instruction in their classrooms. In our early conversations as a committee, we recognized that we had a common dilemma: traditional mathematics curriculum and instruction had not developed deep mathematical understanding in our students, and our practice of filling the class period with hands-on activities also fell short of providing the desired mathematical understanding. This became more apparent as we began to look at the learning of mathematics in our district from a K–12 perspective—concepts that teachers were convinced students had mastered at a lower grade seemed like foreign ideas when encountered again in new contexts in higher grades. This lack of retention of concepts bothered us as a committee, and we recognized that we were spending a great portion of our mathematical instruction time repeating and revisiting ideas that should have been mastered previously.

In search of solutions, our committee read and discussed the draft National Council of Teachers of Mathematics [NCTM] *Standards 2000*, which would eventually be published as *Principles and Standards for School Mathematics* (NCTM, 2000). At the same time, several members of our committee who had become acquainted with a professor at a local university involved in conducting her own research on how children learn mathematics recommended that committee members enroll in a two-semester course she was teaching on the subject. While the focus of the course was on elementary school children, Sharon, Scott, and Vicki, a high school member of the alignment committee, decided to enroll. We wanted to show our support of the elementary school teachers in our district and to deepen our own understanding of the work they were doing to change the nature of mathematics instruction at the elementary level. Little did we realize what a profound impact this knowledge about early childhood research would have on our own classroom instruction at the secondary level, and on our growing understanding of the central message of the NCTM *Standards*—teaching mathematics for understanding.

During these two semesters we had our first in-depth encounter with research on how elementary school children learn mathematics. As a class, we read and discussed manuscripts such as *Children's Mathematics: Cognitively Guided Instruction* (Carpenter, Fennema, Franke, Levi, & Empson, 1999); *Developing Mathematical Ideas: Building a System of Tens* (Schifter, Bastable, & Russell, 1999); *Developing Mathematical Ideas: Making Meaning of Operations* (Schifter, Bastable, & Russell, 1999); and *Thinking Mathematically: Integrating Arithmetic and Algebra in Elementary School* (Carpenter, Franke, & Levi, 2003). As we read these works, we began to reflect on how our own mathematical conceptions (and misconceptions) had evolved.

Over time, those of us in the class who were secondary school teachers began to see connections between the mathematical learning of elementary school children and that of our students. We noticed connections between the way children develop mathematical concepts at an early age, and what they struggle with as secondary school mathematics students. As secondary mathematics teachers we were unaware of the strategies students use to build mathematical ideas—we realized that we expected rote repetition of abstract skills and procedures, unaware of the multiple strategies students might use to solve a problem—and we were not adept at listening to student thinking as they grapple with problem situations and conceptual ideas. We began to see the power of extending pedagogical techniques from the elementary school—such as using rich story-contexts to build students' conceptual understanding of mathematics—to the secondary school. As part of our work, we investigated student understanding of mathematics at the secondary level based on extensions of ideas from research at the elementary level. In this chapter we explore three insights that emerged from our investigations:

1. Mathematically rich stories allow students to construct mathematical procedures and strategies as a by-product of their effort spent working on a problem;

2. Student-written stories elicit evidence of their misconceptions about mathematics as well as indicating their current depth of understanding;

3. Conversations about mathematics allow students to construct meaning for big conceptual ideas.

These insights are elaborated on in the sections that follow.

THE POWER OF MATHEMATICALLY RICH STORIES: INVENTING PROCEDURES THROUGH PROBLEMS

During one of our university sessions we were given a set of addition and subtraction story problems and were asked to organize them according to level of difficulty. The secondary school teachers in the class struggled with seeing one problem as more difficult than another, since we reduced each problem to an equation that could easily be solved. The elementary teachers in the group, however, recognized that children would perceive each of the problems differently because they would focus more on the action of the story rather than the abstract equations that we were writing to translate the story into symbols. This was a new insight for the secondary teachers—to see problems as actions to be modeled rather than words to be translated into abstract symbols. Consider the following as an example: "Connie had some marbles. Juan gave her 5 more. Now she has 13 marbles. How many marbles did Connie have to start with?" (Carpenter et al., 1999). Secondary school teachers solved this problem by subtracting 5 from 13, subconsciously representing the problem with the equation $x + 5 = 13$ and its solution $x = 13 - 5$. However, the research of Carpenter et al. (1999) indicates that children focus on the joining action of the story, making this an addition problem with an unknown starting amount.

Initially, young children might model this problem by starting with an unknown amount of cubes, adding 5 more to the set, and then counting all of the cubes to see if they have 13. If not, they would adjust the initial handful of cubes, adding to it or taking some away, until they have the correct total. This modeling strategy often leads students to recognize that they can start with the known amount of 5 cubes and count on until they have a total of 13 cubes, while keeping track of the additional cubes they have added. As their number sense evolves, students might use landmark numbers, such as 10, to derive more efficient strategies for counting to 13: "I have 5 objects, 5 more would make 10, and then 3 more would make 13, so I've added 5 and 3 more, or 8 objects altogether." Ultimately, they would see 13 as the sum of 5 and 8 as a known fact.

In *Children's Mathematics: Cognitively Guided Instruction*, Carpenter et al. (1999) effectively describe how children's initial thinking about mathematics is based on their ability to directly model the action in a story. Direct modeling strategies evolve into counting strategies and the use of derived facts as children start to identify essential elements of a story that are common to other similar stories. This leads to natural, and yet sophisticated, strategies for dealing with more and more difficult problem types. Stories, it seems, are the ideal mathematical tasks to allow students to develop computational skills with understanding.

We were impressed with how powerfully elementary students use the action in a story to develop their own mathematical strategies without being given specific rules or procedures to follow. As we reflected on the role of stories in teaching computation, we began to wonder about the role stories could play in the development of algebraic thinking and symbolic manipulation skills. In analyzing research at the elementary level, we identified four key contributions that stories can make to the teaching and learning of computation and hypothesized that they may provide parallel benefits for the teaching and learning of algebra (see Table 8.1).

We wondered if secondary school students could also develop their own strategies if given appropriate mathematical tasks. This natural process of moving from direct modeling to sophisticated strategies had generally not been available to our secondary school mathematics students. Rather than starting with stories that provide actions that can be modeled mathematically, these students generally first encountered rules and procedures that lacked a context that could provide meaning for the mathematics they were learning.

As part of the district's move to adopt *Standards*-based mathematics curricula, Scott was piloting a curriculum unit titled "High Dive" from the *Interactive Mathematics Program* (Fendel, Resek, Alper, and Fraser 2000) in his precalculus honors classes. This gave our research group the opportunity to test out some of our hypotheses about the use of stories in developing algebraic thinking. "High Dive" uses the story of a circus act, in which a diver will be dropped from a moving Ferris wheel into a moving cart of water, to help students create meaning for circular, linear, and parabolic motion.

Table 8.1.   Extending the Research on the Role of Stories

Stories at the elementary level provide:	Stories in algebra may provide:
• a pathway for stepping through the problem in a systematic way by thinking through the events in the story	• a pathway for identifying patterns which could then be generalized with symbols
• an action, which would allow students to sort out what operations they need to use (e.g., joining, separating, grouping, partitioning);	• an action, which would allow students to sort out what operations, relations, and functions they need to use (e.g., constant rates of change, growth factors, periodic behavior, etc.)
• multiple entry points into the problem (e.g., direct modeling, counting strategies, derived fact strategies, fact knowledge, and invented algorithms)	• multiple entry points into the problem (e.g., direct modeling, indirect modeling, derived strategies, and invented algorithms)
• schema to relate the physical objects in the story to the numerical representations	• schema to relate the physical objects or scenarios in the story to their algebraic representations

On Day 1 of the "High Dive" unit, students were trying to determine the height a person would be above the ground at different positions on the Ferris wheel, using only their knowledge of right triangle trigonometry. By Day 2, they were writing equations for a person's height above the ground as a function of the time they had been riding on the Ferris wheel. By Day 3, they were modifying their equations to represent rides on other wheels with different diameters, rotational periods, or heights of the center of the wheel above the ground. Concepts such as amplitude, period, and horizontal shift—and the parameters in a trigonometric function that affect these quantities—were natural extensions of the story of the Ferris wheel, and Scott found that these students did not have the conceptual struggles typical of his past students who had been taught about trigonometric graphs in a more traditional way.

The real power of the Ferris wheel metaphor became apparent when the students encountered a new context on Day 4 of the unit: building sand castles at the beach while paying attention to the periodic behavior of the tides. They encountered this new scenario in a homework assignment, without the support of additional instruction or group conversation. The next day several students wanted to share their work. "The tides are like the Ferris wheel," Doug said, "only we need to make the following changes in our thinking."

The Ferris wheel has become a metaphor for Scott's students to use to think about all types of periodic behavior. The students had no memorized rules for determining the values of the parameters representing period, amplitude, or phase shift—just their stories that began "if this were a Ferris wheel"—but they were much more accurate than past students in sketching trigonometric graphs and translating periodic behavior into trigonometric equations.

When implementing a complex problem such as "High Dive," it is often tempting to be very explicit about what students should do in each step along the way. That is not the authors' intent for this curriculum, however. Rather, the individual tasks throughout the unit are designed to develop conceptual ideas and strategic ways of thinking about topics like periodic behavior and how such behavior can be modeled with trigonometric functions. Scott's knowledge of the results obtained by elementary school children who encountered these same types of rich mathematical tasks enabled him to place more faith in the power of the story and the authors' carefully selected tasks to facilitate students' development of procedures than he might have otherwise. As a result, we were able to see that, indeed, stories can be powerful tools for building mathematical skills and procedures.

STUDENT WRITTEN STORIES:
REVEALING DEPTH OF UNDERSTANDING

During one of our university class sessions, our instructor forced us to encounter our own misconceptions about mathematics by asking us to write a story and a question which would be answered by the arithmetic sentence $2\frac{3}{4} \div \frac{1}{2}$. We have since used this same activity with numerous high school students and adult study groups. The majority of stories that high school students and adults write take on the form of sharing something—such as $2\frac{3}{4}$ pizzas—equally between two people, failing to note that such stories fit the equation $2\frac{3}{4} \div 2$. Occasionally, someone will think about the arithmetic of "invert and multiply" that they were taught to use in elementary school and will write a story to fit $2\frac{3}{4} \times 2$; for example, "A recipe calls for $2\frac{3}{4}$ cups of flour and you want to double the batch, so how many cups do you need?" The writers of such stories seem to think that this multiplication story is equivalent to the original division situation. Only on very rare occasions will someone identify this problem as a measurement division and write an appropriate story, such as a recipe calls for $2\frac{3}{4}$ cup of flour, but I can only find my $\frac{1}{2}$ cup scoop—how many scoops do I need? We were fascinated with this inability to tell a story to accompany a very familiar piece of arithmetic symbolism, so we decided to investigate what our students understood about the symbolic notation of algebraic expressions and equations.

We began with Sharon's students, who we asked to write summary phrases to describe the meaning of algebraic expressions made from predefined variables. This task was designed to find out what students understand about the use of variables in mathematical expressions—what the variables mean and what combinations of variables and operations make sense. We felt this task would give us a good sense of what students understand about the abstract symbolic notation they use in algebra. For example, if B represents the number of boys in a class, G represents the number of girls, and L represents the cost of a lunch for each student, then $L(B + G)$ represents the amount the students in the class spent for lunch. However, given the expression LS, where L represents the cost of a lunch and S represents the cost of a snack, the majority of her students wrote "the total amount one student spent for food" as their summary phrase, even though students could clearly explain that the notation represented multiplication of the two variables. The students failed to recognize that the expression LS had no meaning in this context, and they merely made up a summary phrase that accounted for the variables, but not for the action represented by the expression.

We found that this inability to account for the action in an expression or equation persisted into advanced algebra. For example, Adrianne, another high school member of the alignment committee, was amazed when the students in her Algebra 2 honors class were unable to write a story that would be illustrated by the equation $50t - 200 = 125$. Joseph wrote, "Mrs. Jackson was giving some pieces of paper to each of the 250 students in her classes. 200 students were absent. In the end she had only given out 125 full sheets of paper. How many sheets of paper did each of the remaining 50 students get?" In this analysis, Joseph does not understand that $50t$ represents the total amount of paper distributed (therefore, $50t = 125$), nor does he recognize that he is subtracting 200 students from $50t$ pieces of paper.

Students often used the steps involved in solving an equation to create a story for the equation, just as the students and adults who know that in order to solve $2\frac{3}{4} \div \frac{1}{2}$ they need to invert and multiply will create a story that requires doubling the original amount. For example, Anne wrote the following story for the equation $3(x - 4) = 105$: "You have three groups of friends at your Halloween party. When 10:00 comes around 12 friends leave, giving you a total amount of 105 [remaining]. How many friends were in each group?" Anne's story represents the equation $3x - 12 = 105$. She has already started to solve the equation by distributing the 3 into the $(x - 4)$ factor before creating her story.

Making sense of when to subtract the 4—or what to subtract the 4 from—when writing stories for the equation $3(x - 4) = 105$, was a recurring problem for many students. For example, Sarah wrote the following story: "Tom has 105 jelly beans. He wants to give them to each of his three friends equally. But before he hands out all 105, he decides to keep 4 from each friend. How many do they each get?" Her answer was 39. Sarah's story implies that Tom is taking away three groups of 4 jelly beans from the 105 original jelly beans before he distributes the remaining jelly beans equally among the three friends. The answer to the story that Sarah wrote is 31 jelly beans, while the answer to the equation she was solving is 39.

Kyle wrote this story for the same equation: "You are making apple cider; you want to put it into 3 different jugs. You have 105 ounces of cider; you drink 12 ounces. Now with only that much cider left how many ounces would you put into each jug?" Kyle gave no answer to his problem. Like Anne, Kyle is starting midway in the solution process by distributing the 3, but instead of subtracting 12 from the $3x$, he is subtracting 12 from the 105.

Some students ignored the action of the story completely, merely translating the symbols into words, such as in Leslie's story: "You have an x amount of cd's but you are subtracting 4 from that amount. You have three times more cd's that you want to add in and also subtract from the bunch. All together you have 105 cd's. How many x amount of cd's do you have?"

Only a few students, such as Nate, made sense of the problem. His story reads: "A certain number of people entered a talent show. Each were given three minutes to perform. When the show started, 4 of the people didn't show up. As a result, the talent show only lasted 105 minutes. How many people signed up initially?"

None of these students had problems solving the equation. So why did so many struggle so much when trying to interpret what story the equation might be telling? Do they even know that the symbols could tell a story? How can they judge if they have correctly modeled a problem situation algebraically if they don't see the connections between the symbols and the details of the story? We live in a day and age where technology is increasingly being used to solve equations efficiently and accurately. The algebra skill our students will most need is the ability to tell a story symbolically. But, as we were seeing from their attempts at writing stories to fit equations, this skill is not being developed by traditional algebraic instruction. For example, when asked to create a situation that would appropriately be represented by the equation $x + 6 = 18 - x$, many students wrote something like the following: "John had some money and then found 6 dollars on the ground. His friend had 18 dollars and lost some money. They now both have the same amount." Do students recognize that the x variable on each side of the equation needs to represent the same physical quantity of money and not just some unknown amount? How do we know that John had as much money to begin with as his friend lost? These types of questions lead to interesting conversations as students consider the actions in their stories in relation to the given equation—conversations which most students never had an opportunity to participate in when they were part of a traditional classroom setting. Through the stories that students wrote, we gained evidence of their misconceptions and a more thorough assessment of their current levels of understanding of algebraic symbolism.

CONSTRUCTING MEANING THROUGH CONVERSATION

Initially, our university instructor asked each of us to find elementary mathematics classes to observe and write case studies about. As the course progressed, we asked for and were granted permission to study our own secondary school students' mathematical understanding. By researching topics that were parallel to those in the elementary teachers' studies, we were able to paint a K-12 picture of students struggling with a mathematical concept and the implications of their struggles throughout the grades.

One of the topics we examined was what students understand about equality. For many elementary students the equals sign means, "Here comes the answer." As our class discussed this erroneous notion of equality, we began to wonder if this notion perseveres in secondary school students. Once we began to look, it wasn't hard to find examples of the persistence of this erroneous notion of equality as it constantly showed up in students' written work. One example is the following notation, which appeared on a student's homework paper when he was asked to evaluate the expression for the given value of x:

$$5)\ 6x^2 + x - 11 \text{ when } x = 2$$

$$2^2 = 4 \cdot 6 = 24 + 2 = 26 - 11 = 15$$

$$5)\ 6x^2 + x - 11 \text{ when } x = 2$$

$$(2 \times 2)\ (2 - 11)$$
$$(6 \times 4)\qquad -9$$
$$24 + -9$$
$$15$$

In this example we see Tyson using the equals sign as if it were just an indicator that he was moving on to the next stage of the evaluation process. He ignored appropriate uses of equality, introducing his own misconceptions in his notation. Students are often encouraged to show how they are thinking about a problem, but this type of representation reinforces their erroneous notions about equality. On the other hand, here is another student's representation of her thinking about this same problem where the notion of equality was not an issue. Instead, this student's work focuses on the substitution and evaluation process.

We wondered how the erroneous notion of the equals sign as an indicator of action might hinder algebra students' ability to learn how to solve equations where the notion of "balance" is essential to their understanding. To investigate this issue, Vicki and Scott received permission to conduct the following study with a beginning algebra class at their high school. Of the 14 students in this class, 11 were sophomores, 2 were juniors, and 1 was a senior. Most students in our district complete beginning algebra in junior high, thus students in this class are typically those who struggle with mathematical ideas. They have received traditional mathematics instruction during their entire school experience and have not been successful. The study was conducted in November after the students had completed a unit on solving one- and two-step equations. Vicki and Scott interviewed the classroom teacher and determined that the instruction the students had received on solving equations had been a typical, procedurally-driven approach.

Each of the students in the class was given 10 pieces of paper on which to record their name, answers, and any calculations they used to obtain their answers. Questions were posted one at a time for all of the students to see, and their answers were collected after each question was completed. Questions were presented in sequence and no discussion about the questions was allowed. Vicki and Scott used this first session with the beginning algebra students merely to collect data about their understanding of equality.

The questions (see Figure 8.1) were designed to elicit information about student thinking on each of the four benchmarks of equality defined by Carpenter et al. (2003):

Benchmark 1:   Understanding of the equal sign ("balance" vs. "here comes the answer");

Benchmark 2:   Understanding alternative formats involving equality (e.g., the answer can precede the equal sign);

Benchmark 3:   Recognize equality as equivalence (balance); compare by calculating each side; and

Benchmark 4:   Recognize equality as equivalence (balance); use relational thinking to compare each side.

Benchmark 1: Understanding of the Equal Sign

We had not anticipated the depth of misconceptions of equality that existed among these high school students. While only 2 students interpreted the equal sign as "here comes the answer" on question #1 by filling in the box with the number 100, 7 students added all three numbers together and filled in the box with 157. On question #4, 7 students accumulated the answers as they moved from expression to expression (e.g., $32 + 48 = 80 + 15 = 95$). Likewise, 9 students incorrectly stated that question #10 was a true compound equality statement. Only 1 student commented that question #10 was like a "run-on sentence." It is apparent that the majority of students in this class have not distinguished between the equal sign as indicating "balance" versus "here comes the answer," and they are unclear as to what numbers constitute the answer.

Equality Interview Questions

Fill in the box with the correct number.

Example: 20 + 80 = ☐

1. 22 + 78 = ☐ + 57
2. 716 + 213 = 714 + ☐
3. 22 x 13 = 11 x ☐
4. 32 + 48 = ☐ + 15 = ○

Determine if the following statements are true or false.

5. 75 = 40 + 35
6. 5 + ☐ = ☐ + 5
7. 325 + 114 = 320 + 119
8. 14 x 8 = 7 x 16
9. 12 x 46 = 24 x 23
10. 3 + 7 = 10 + 5 = 15 + 2 = 17

Figure 8.1.   Equality Interview Questions.

Benchmark 2: Understanding Alternative Formats Involving Equality

Most students seemed flexible with alternative formats involving equality, indicating no concern with the answer preceding the equal sign, as in question #5. All 14 students identified this as a true statement.

Benchmarks 3 and 4: Recognizing Equality as Equivalence

It was interesting to note that on question #2 1 student added all of the numbers together, 1 student answered incorrectly with no calculation, 1 student unsuccessfully tried to use algebraic properties to solve for ☐ , and 1 student wrote that the problem made no sense. Five students found the correct answer of 215 by summing the numbers on one side and then subtracting 714 to get their answer. Only 4 students used relational thinking and got the correct answer without calculations. One student attempted to use relational thinking, but decreased the number to 211 when he should have increased the number to 215 in order to maintain the balance.

Questions #3, #8, and #9 were intended to address relational thinking with multiplication. As we tallied the results for question #3, we noted that 1 student multiplied all three numbers together. Ten others tried to use algebraic properties to solve for □, although only 3 were successful. Again only 4 students used relational thinking and correctly identified the missing factor as being twice as big since the first factor had been cut in half.

To use these questions to gain additional insight into the students, we examined the collective set of responses from each individual student. Hilary is an example of how an individual student responded to the equality study questions.

Hilary is a sophomore in this class. She does not see equality as "balance." On question #1 she added all three numbers together. On #4 she filled in the□ and the ◯ with the answers to the calculations on the left of the equal sign. She believes that #10 is a true statement. Therefore, it seems that Hilary has not developed the conception of the equal sign as defined in Benchmark 1. She has achieved Benchmark 3 as indicated by questions #7, #8, and #9, where she correctly calculated the value of the expressions on each side of the equal sign to determine equality. She demonstrated no relational thinking—she always used calculation to compare the two expressions.

In *Thinking Mathematically*, Carpenter et al. (2003) reported that children clarify their notions of equality when they are put in a position that challenges their existing conceptions. They suggest that this can best be accomplished by "engaging them in discussions in which different conceptions of the equal sign emerge and must be resolved" (p. 14). It appeared that the traditional instruction for solving equations that students in our study had received had not sufficiently challenged their conceptions, and thus had not overcome their misconceptions of equality. After discussing this research in our university class, Vicki and Scott decided to return to the beginning algebra class to have a follow-up conversation on equality and relational thinking. We began by sharing the responses of the students to the question 22 + 78 =□ + 57. Previous responses to this question were 157 (7 students), 43 (5 students), and 100 (2 students). We asked if the students could see how the three different answers were obtained. We also asked them if they felt that it was okay for the problem to have three different answers. The students could explain how all of the answers had been obtained, and most were okay with considering all of the answers as being correct. Some students thought that only one answer should be correct, but there was no consensus on which one of the three answers that should be.

We then presented the students with the following sequence of true/false statements:

$$22 + 78 = 22 + 78$$

$$22 + 78 = 78 + 22$$

$$22 + 78 = 24 + 76$$

All the students agreed that the first two statements were true. Apparently they had abandoned the "here comes the answer" paradigm in these situations, since no one wanted to argue that the statements were false because the answer, 100, didn't appear immediately following the equals sign. However, the question still remained: Do the students see the two sides as being equal because they involve the same amount of "stuff" on both sides, e.g., the sides are in balance, or do they see the two sides as being equal because the right side is just a restatement of the problem on the left? The fact that several students disagreed with the third statement suggests that the dissenting students saw equality as allowing the addition problem to be restated using exactly the same two numbers added together on each side of the equal sign, as in equation #2, but that they did not see equality as balance, such as when the numbers added together were not the same, even though the sum on the right was equivalent to the sum on the left, as in equation #3. When asked to explain why he thought equation #3 was true, one student used a football analogy. "If you are on the 22 yard line and you make a 78 yard run, it is the same as if you start on the 24 yard line and run 76 yards." Another student agreed, "It's like you lowered one number and raised the other number by the same amount." This seemed to convince most students that the third statement was true.

We extended our list of true/false statements to include:

$$22 + 78 = 32 + 68$$

$$22 + 78 = 42 + 58$$

We incremented the first number by 10 and decremented the second number by 10 so that students could follow the same line of reasoning as presented in the previous student explanations. One student added the analogy that "it's like dollars and change, you can make $100 in different ways by exchanging $10 bills from one pile to another."

We then presented the original problem again: $22 + 78 = \square + 57$. This time one of the students immediately remarked, "It has to be 43." But not everyone was willing to agree. The student with the money analogy explained, "It's like we're going on a trip and I pitched in $22 and then later pitched in $78 and my friend pitched in $57 and needs to pitch in $43 more." He recognized the balance nature of the equals sign and saw the two sides of the equation as two different ways to express a relationship among numbers.

To see if the idea of "balance" of equality would transfer to statements involving multiplication, we reintroduced the following statement from our equality questions.

$$22 \times 13 = 11 \times \square$$

Right away, several of the students could see that the number in the box should be 26. They used relational thinking to explain, "You cut the first number in half so you need to double the second number."

We also reintroduced the true/false statements:

$$14 \times 8 = 7 \times 16$$

$$12 \times 46 = 24 \times 23$$

Whereas previously we saw most of the students multiplying the quantities on each side of the equation to determine if the statements were true or false, this time many readily responded that both statements were true, without using paper and pencil to compute their answers. The relational thinking of doubling one factor while cutting the other factor in half became available to students once that had recognized that equality represents "balance."

By the end of the session, all but three students accepted this "balance" idea of equality and, therefore, could use relational thinking to balance the sides of the equation. Perhaps what impressed us the most was that this change of perspective had come about through the students' own conversations about the mathematics. No one was telling the students what to believe; rather, they were constructing their own beliefs about the nature of equality through engagement with the tasks and carefully chosen follow-up questions.

CONCLUSIONS

As we think about how students learn from their earliest experiences with mathematics through their high school experiences, we need to consider what misconceptions they are acquiring and maintaining along the way. Through examining the research on how elementary school children learn mathematics, we have acquired tools that we use in our secondary school classrooms—both to help students examine and overcome some of their misconceptions and to extend their understanding of mathematical ideas. These tools involve using stories to develop the mathematics, rather than just to apply the mathematics; having students write mathematical stories that provide a window into their understanding and misconceptions about symbolic notation; and using classroom conversations to debate the meaning of mathematical symbols and concepts. Of course, the success of all of these tools is dependent on the teachers' willingness to listen to students as they construct their own understandings about mathematics, and then to ask probing questions that help students revise and refine their own thinking about big mathematical ideas.

Each of these tools became part of our teaching repertoire after we had seen evidence of their profound effect on young children's thinking about mathematics. As a result, we also modified the professional development approach for secondary school mathematics teachers in our district to include a review of the research on children's mathematics cited in this chapter. We encourage teachers to integrate the recommendations of this research in their instruction through three key actions: (1) begin instruction with a story—a contextualized problem situation that provides a pathway for thinking about new mathematical concepts and solution strategies; (2) facilitate conversations in which students share their solutions strategies and discuss the meaning of the symbols they use to represent their thinking; and (3) have students reflect on how the symbols they use tell a story and how the actions in a story relate to the mathematics. These ideas are easily illustrated by examples in the research on children's mathematical thinking, and once these principles are recognized, they can be more easily adapted to the secondary level. We also encourage teachers to conduct their own informal research in their classrooms. We have found that listening to students' thinking—their developing ideas and misconceptions, their invented algorithms and representations—facilitates allowing their mathematical thinking to evolve in a natural way. We have learned to trust this methodology because of our introduction to the research on children's mathematical thinking and the results we have seen when we have applied it to our own classrooms.

In addition, as secondary teachers we have been surprised by what we thought our students knew, and how we thought their knowledge about mathematics had been acquired. It became apparent that mathematics instruction needs to be viewed from a K–12 perspective, and not as isolated grades and courses. One of the benefits to our district from adopting standards-based curricula has been the rich conversations that have flowed across grade levels as we have tried to make sense of the research on mathematics instruction. Secondary teachers have become more adept at looking at mathematics from a student's perspective and allowing a variety of strategies to be developed and discussed, while elementary teachers have become more aware of how the big ideas of computation and number sense lay a foundation for algebraic thinking. In the process of implementing new curricula—and throughout the discussions surrounding this implementation—the beliefs and practices of both secondary and elementary mathematics teachers have changed to the benefit of our students.

NOTE

1. Scott was at Lone Peak High School at the time of the research reported in this paper.

REFERENCES

Carpenter, T. P., Fennema, E., Franke, M. L., Levi, L., & Empson, S. B. (1999). *Children's mathematics: Cognitively guided instruction*. Portsmouth, NH: Heinemann.

Carpenter, T. P., Franke, M. L., & Levi, L. (2003). *Thinking mathematically: Integrating arithmetic and algebra in elementary school*. Portsmouth, NH: Heinemann.

Fendel, D., Resek, D., Alper, L., & Fraser, S. (2000). *Interactive Mathematics Program, Year 4*. Emeryville, CA: Key Curriculum Press.

National Council of Teachers of Mathematics. (1989). *Curriculum and evaluation standards for school mathematics*. Reston, VA: Author.

National Council of Teachers of Mathematics. (1991). *Professional standards for school mathematics*. Reston, VA: Author.

National Council of Teachers of Mathematics. (1995). *Assessment standards for school mathematics*. Reston, VA: Author.

National Council of Teachers of Mathematics. (2000). *Principles and standards for school mathematics*. Reston, VA: Author.

Schifter, D., Bastable, V., & Russell, S. J. (1999). *Developing mathematical ideas: Building a system of tens*. Parsippany, NJ: Dale Seymour.

Schifter, D., Bastable, V., & Russell, S. J. (1999). *Developing mathematical ideas: Making meaning of operations*. Parsippany, NJ: Dale Seymour.

CHAPTER 9

GIVING VOICE TO SUCCESS IN MATHEMATICS CLASS

P. Janelle McFeetors
River East Collegiate

Student achievement in high school mathematics has become the focus of many discussions in mathematics education. Factors such as high school exit examinations, international standardized tests, workplace preparation needs, and an increasingly technological society have raised awareness of student progress. Students, teachers, administration, parents, and the general public are concerned with student achievement in mathematics and, more specifically, with increasing student achievement. Much of the public focus is on examination scores as an indicator of achievement in high school mathematics. Defining success in terms of examination scores overlooks other aspects of success that may have more meaning to the teachers and students who live out the tensions of teaching and learning each day within a classroom.

As a high school mathematics teacher, I decided to listen to my students as a way to better understand the multidimensional nature of their success. The following excerpt[1] captures the success of Erin, a student in my Consumer Mathematics class, as she narrated it in her end-of-semester portfolio:

Teachers Engaged in Research
Inquiry Into Mathematics Classrooms, Grades 9–12, pages 153–174
Copyright © 2006 by Information Age Publishing

I improved in learning math over this semester by practicing more and especially with the portfolios because it lets you look over and go over them assignments and have another look at them. I learned how to take my time and have patience with a question and how to go over it and keep trying and making up little tricks in your head to get a better knowledge of it and to get the correct answer in the end. Another thing that I kind of got hooked on was showing all the work even when you had a simple way of just jotting it down because if you make a mistake then you can get half marks at least for trying and maybe you could even get the first half the question right and just did some calculations wrong but you let the teacher see that you know what you are doing but maybe just wasn't paying to much attention when you are finding the final answer.

Some learning strategies were making up little tricks in my head to help me remember a certain way to do a question . . . practicing over and over again till you get the hang of it . . . go over my notes briefly . . . I'm going to use lots of these strategies in my next math course because it is the exact same thing just at a high level and at a next step. . . . I liked this coarse a lot and I think that it is going to do a lot for me in the future because I got the hang of everything pretty good. Lots of my strategies were successful and I will use them I just hope that I can remember them, but most of them were from my head so I think if I look at a specific question for long enough and try it out a couple times I will remember how to do it.

Although marks and practicing questions were still important to Erin, she also included learning about how she was thinking and learning as her moments of success—the capstone of her success in Consumer Mathematics. Erin did not begin the semester with confidence in her ability to learn mathematics or with the ability to be explicit about her success. As her teacher, I wanted to both foster success for Erin and come to understand the nature of her success and how it evolved over a semester. I anticipated that her success, and that of her classmates, might be diverse (not only related to mathematical cognition), tentative, and occur within our classroom processes. The complexity of noticing success highlights the importance of inquiring within a classroom setting where the teacher lives in pedagogical relationship (van Manen, 1986) with the learners. For me, this meant designing an inquiry with students whom I was teaching, so that the authenticity of my questions, the students' experiences, and my coming to understand would respect the complexity of teaching and learning.

The purpose of this chapter is multifaceted, to reflect the complexity I experienced in inquiring into my own practice and listening intently to the students I was teaching. Erin's story of success provides the framework for the chapter. To capture a student and his or her story is a challenging endeavor; I have highlighted critical moments in Erin's story that portray the complexity of coming to succeed and the complexity of a classroom-based inquiry. Rather than relying on other research studies or assumptions from previous experiences, as an inquirer I focused on an emerging understanding of success—grounded in the students' experiences and explicated through conversational opportunities, the students and I negotiated a sense of success that is described in this chapter. The nature of success was not (pre)defined by me nor was it defined explicitly by the students; it emerged through experiences and conversations reflectively, thoughtfully, (pro)actively, and interactively. The success of each student was fostered by a listening-based stance, thus the re-telling of Erin's story also provides examples of how a mathematics teacher can listen to her students.

INQUIRING INTO SUCCESS:
UNDERSTANDING THROUGH STORIES

Erin and her classmates were enrolled in a grade 10 Consumer Mathematics course. In Manitoba, Canada, students select from among three mathematics strands in order to satisfy the requirements of obtaining a mathematics credit in each year of high school. Consumer Mathematics focuses on using reasoning to support wise financial decisions and prepares students to enter the workforce directly after high school, while still meeting entrance requirements of some university programs. The class was composed of students in grade 10 and 11 who had chosen the course because they believed they did not have the ability to learn mathematics and had come to dislike mathematics and mathematics class. These students had been effectively marginalized by their previous experiences in mathematics class. Eleven of the students (approximately half the class) volunteered to participate in the inquiry I conducted.

I decided to situate an inquiry into success in a Consumer Mathematics class because I had noticed that the students' sense of accomplishment and success at the end of a semester did not come directly from test scores. Success for these students seemed to come from a shift in stance toward mathematics and mathematics class. Through my on-going inquiry in my classroom, I had come to see that two factors fostered students' progress, namely teaching as a relational act in which the teacher cares for her learners (Noddings, 1984; van Manen, 1986), and a curriculum that provides opportunities to understand mathematical concepts (Manitoba Education, Training and Youth [METY], 2002). The curriculum for Consumer Mathematics highlights connections between mathematics and the world and contains a focus on "assessment . . . to guide and enhance [student] learning" (National Council of Teachers of Mathematics [NCTM], 2000, p. 22). As I considered my learning goals within the inquiry, I recognized that I wanted to understand the lived, yet barely articulated stories of learners as they experience their first successes in high school mathematics.

The above narrative text that Erin authored was part of the data that I collected in my inquiry. Collecting narrative texts and using stories to interpret the experiences of Erin and her classmates was central to my research methodology, narrative inquiry (Clandinin & Connelly, 2000). Using listening as a methodological framing necessitated adapting narrative inquiry so that the voices of the students could be heard in order to draw meaning from their experiences. Although I collected and analyzed data from all 11 students who agreed to participate, this chapter tells the story of Erin. As Erin authored interactive writings (17 in total) and created unit portfolios (5 in total), she was beginning to tell me about her success in mathematics as she noticed and understood it. The interactive writings (Mason & McFeetors, 2002) prompted Erin to reflect on her learning and progress in the course, and I in turn responded to her and the ideas she raised. Unit portfolios were summative unit assessments for which Erin selected six items and described what learning and mathematical processes were evident in each, along with authoring her learning story for the unit in an overview reflection. Erin's description of her success in mathematics class through these classroom processes is consistent with the idea that "students learn more and learn better when they can take control of their learning" (NCTM, 2000, p. 21).

Erin's awareness of her own success in mathematics class was also supported by three narratives of success I wrote about her and our subsequent conversations (three informal interviews) about her success. Each narrative retold several moments of the students' success that I noticed in their data pieces or my daily field notes. Central to the narrative was my wondering about the meaning of events within a specific theme of success. Together, these five types of data, within the inquiry viewed as data collection and interpretation, provided opportunities for me to listen intently to Erin and her classmates and come to understand the nature of their success in Consumer Mathematics. My analysis of the data was both ongoing as I used the data to inform my teaching and reflective as I revisited the completed data set after the semester had ended to identify themes and insights that might not have been apparent when I was focused on my students' learning.

ERIN'S STORY OF SUCCESS:
THINKING LIKE A MATH LEARNER AND A MATH THINKER

Erin was a grade 11 student who found herself in a position of catching up with her grade 10 mathematics credit because she had dropped out of school the previous year. She knew several of the grade 11 students in the class and was outgoing. She interacted with her table partners and often found other students to learn with as well. Erin was intent on staying in school and finishing her grade 10 credits, so she took seriously doing well in the course and maintaining her attendance. This was evident in her focus and work habits in class, and her responsibility in completing assigned tasks. Although she wrote in her first interactive writing, "I'm bad at math," throughout the semester Erin indicated to me an ability to think mathematically.

Erin is an exemplar of the success that the learners in this inquiry experienced. An exemplar is an individual who serves as an ideal example for a group of individuals. Erin is an exemplar because her story contains elements of success that I noticed in all of the learners, not because she experienced the most or least significant success. Each of the learners had a particular success, connected closely to his or her experiences, highlighting the unique nature of each learner's journey. Erin's story exemplifies, on behalf of all her classmates, that being successful was a tentative process and one in which I needed to listen closely to their moments of success to catch the subtle shifts in stance. These subtle shifts in stance contributed to their (re)forming of identity—in essence, living a dynamic process of *becoming*[2]. Success for each of the learners was not an isolated moment, but a journey as they were forming their identity in mathematics class as students, learners, cognizers, and human beings in relation with others. Erin's story illustrates this evolving sense of success.

An Initial Stance: A Limited Discourse

The following segment of Erin's story demonstrates a progression that occurred with many of the learners' experiences, the data collected, and my interim data interpretation. Not only does a sense of the flow of data become apparent, but also the stance with which Erin began the semester in Consumer Mathematics. As with all the learners involved in the inquiry, I was challenged as an inquirer to initiate conversations with Erin about what she believed about mathematics, mathematical thinking, learning, herself, and her success at the beginning of the semester. It was challenging to listen to Erin in the little that she expressed to me, a situation that was similar with her classmates.

About two months into the course, I invited the students to notice counting patterns with fractal cards (Simmt & Davis, 1998). Erin was working with her table partner, Karl, to notice and then express the growth of several parts of their fractal card. In my field notes, I recorded:

> I was helping Erin and Karl at one point in the class. Erin was asking a question and got a little frustrated when I backed up and started asking a few other questions. I think she felt like it had nothing to do with what she was doing. It did, but I guess she didn't quite see where I was going.

Rather than providing a direct answer to Erin's request for help, I asked guiding questions so Erin could engage in noticing and expressing the pattern herself. The next day in class, Erin and I wrote about the fractal cards in her eighth interactive writing, where she began by writing, "The patterns were mathematical because you had to use math to figure out what exactly the pattern was." I responded by writing, "Erin, I like the idea of the patterns being the mathematical part. But, you also identified the finding of the patterns as being mathematical." Although Erin wrote about the fractal cards as requested, she did not report my interaction with her and Karl, her mathematical thinking, or the pattern she had noticed. I struggled, as an inquirer, to understand how the mathematical task had affected Erin's view of learning mathematics. In the absence of more description, I was challenged to make meaning of what Erin had said.

In preparing for our second conversation, just over a month later, I authored a narrative of success for Erin. For each student, the narrative highlighted several specific moments of success I had noticed, recounting the event and wondering what it meant as the student's success in mathematics class. During the ensuing conversation, I would read one paragraph at a time and then the student and I would discuss it—as I attempted to confirm my interpretation of the student's experiences and invite him or her to describe similar examples of success. The focus of Erin's second narrative was her success in constructing her own algorithms to complete questions and thinking mathematically. I included the fractal card interaction in the narrative because I was not sure how to interpret it as a success. In order to begin the process of interpretation and engage Erin in talking about her successes, I wrote in her narrative:

> Sometimes I have noticed that you would like me to just tell you how to do a question. I remember you demonstrating some frustration when working on fractal cards. When you called me over to help, I decided to guide you through the process instead of giving you the exact steps. I'm not sure how this fits in with your story of success, and I'd like to hear more about it from you.

As this segment of narrative indicates, Erin demonstrated (but did not vocalize) a desire to be told what to do in our interaction. Erin did not communicate verbally to me her frustration during our interaction in class, or why she was frustrated. The narrative provided an opportunity for Erin to interpret the fractal card situation and respond to my wondering, as can be seen in this excerpt from our conversation:

J: So, tell me about that idea of you getting a little frustrated when I don't tell you exactly what the steps are and you have to figure out some of the steps on your own.

E: I couldn't figure that stuff out. I didn't understand how you'd get, like, you'd say "Well, stage 6, how do you get to stage 50, or something." I don't know. I don't know how to do that.

J: Okay. How does that fit in with making your own tricks? Is that a point where you decide? {pause}

E: I didn't understand {inaudible words}

J: You didn't understand, so you couldn't. So to make your tricks, then, do you have to know a little bit about the topic first?

E: Mm, hmm.

      …

J: Why do you get frustrated when I help you through a problem, rather than telling you the answer right away?

E: When I have a question?

J: Yeah.

E: {little laugh} 'Cause I've probably been working for a while and I'm frustrated with the question already.

J: Okay, so you really give it a try before you put up your hand or call me over?

E: Yep.

J: So at that point you want the exact steps to follow?

E: Yeah.

Even within the prompts, which were sometimes quite directive, Erin said very little about the source of her frustration. Notice that when Erin was making a statement that was important or about herself and her learning, she trailed off and spoke so quietly that it was almost inaudible. As an inquirer, it was challenging for me to understand Erin within the absence of words about herself, mathematics, and mathematics class.

At the end of the inquiry, I spent considerable time reading through my data and attempting to come to understand Erin's lack of communication. I had noticed an absence of words not only with Erin, but with her classmates as well, even though they had tried to respond in a genuine manner to the prompts I provided. This contributed to two difficulties for me as an inquirer. First, it was difficult to notice the absence of words while I was teaching the students. It was only as I viewed Erin's words as data, outside of the classroom context, that I came to see the absence. Second, it was difficult to hear what the students were not saying to me. These difficulties limited the kinds of data I received and the sense-making that I could engage in with the data. What I came to notice was that the students' absence of words pointed to a presence of silence in their stance toward mathematics class.

Much of my coming to understand the learners' success in this inquiry was predicated on the reading I had done before and during the study. I sought out as much research related to students' success and ways of expressing that success as was available. The diverse frameworks I found served as a lens to help me make sense of my students' success. Belenky, Clinchy, Goldberger, and Tarule's *Women's Ways of Knowing* (1986) informed my understanding of absence of words as a presence of silence. Silent knowers rely on authority that they perceive to be all-knowing. Their epistemological stance is exemplified in the difficulty they experience describing themselves and engaging in self-reflection. This stance is not an initial stance for individuals, but one that is assumed because of past experiences—for the students in this class a key factor appeared to be their marginalization in previous mathematics classes. Belenky et al. use a metaphor of voicelessness to describe the silence of individuals.

As I began to consider the metaphor of voicelessness to describe the initial voice-stance of Erin and her classmates, I searched for a delineation of voice and examples from the literature. Baxter Magolda's (1992) study with college students provided a starting place. By voice, both Belenky et al. (1986) and Baxter Magolda are referring to a mature, internal voice that defines self-identity, and where the individual develops and expresses his or her own perspective with an authoritative stance. Individuals with mature, internal voice take ownership of their ideas and understanding of self. Returning to the idea of voicelessness, individuals who live in silence do not have a voice—they do not recognize the need to have a voice, they do not focus on gaining a voice (because they do not see the value), and they do not use what little self-awareness they possess to communicate ideas (both mathematical and about themselves).

Erin, and each of her classmates, demonstrated voicelessness at the beginning of the semester. They said very little, or nothing, about themselves, their positioning with authority, their beliefs about the nature of knowledge, their thinking or their learning. Consider several of the examples above, which demonstrate Erin's subordinate positioning with authority. Erin's request to be told the fractal card pattern indicated her belief that her ideas held little value, which is consonant with a voiceless stance. Erin's frustration with saying things about herself is similar to a silent knower's stance, in which he or she believes "that source of self-knowledge is lodged in others—not in the self" (Belenky et al., 1986, p. 31). As well, Erin's use of "I don't know" and trailing off to inaudible words was her way of deferring to the authority's or my knowledge about herself. Belenky et al. described a dependence on authority as a silent knower's perception of "authorities as being all-powerful, if not overpowering" (p. 27).

As Erin's story of success continues to unfold, the concept of voice develops as we come to understand the emergence of voice as the success of these Consumer Mathematics students. Although the challenge to hear students and to notice and articulate their success was a difficult endeavor, their silence formed a backdrop against which the students and I could notice their success. Without the stark contrast of learning in silence, the emergence of voice might not have been heard.

A CHANGING STANCE:
THE BEGINNINGS OF METACOGNITIVE AWARENESS

During the semester, many of the prompts for interactive writings and portfolio reflections asked students to focus on describing their thinking and learning. A few weeks prior to the fractal cards class, Erin had written in her interactive writing after the fourth test that

> I think I'm getting the hang of both rotations and reflections, but if I had to choose one were I thought [I] got more questions wronge I would pick reflections, but I definity understand them. They are kinda tricky.

Erin based her understanding of her thinking on the score given by the teacher, rather than her own analysis of her thinking—a silent knower's stance.

Soon after, I noticed a change in Erin's responses. She began to explicitly respond to the metacognitive elements of the prompts. As a second part of her eighth interactive writing, Erin was asked to describe her thinking for a specific question on the fifth test. The question required students to compare the unit price of 1-, 2-, and 3-packaged t-shirts, in order to find the best price to buy 7 t-shirts. Erin wrote, "On the T-shirt question I just got all the prices and then compared them to see which one was lowest, it was a pretty easy question." By describing her steps, Erin was reporting what she was thinking as she completed a test question. This response was Erin's early attempt at being explicit about some of her thinking. With this interactive writing in mind, I continued to notice when Erin was writing about her thinking. A couple of weeks later, Erin wrote the following reflections for her Spatial Geometry Portfolio:

Item 1:   This is my Reversing [Rod] analogy puzzles. This demonstrates problem solving because your trying to find the rotation or reflection. Also this demonstrates visualization because it is an object. I did some good thinking because you have to figure out if you flipped it the right way.

Item 2:   This is my Spatial Geometry assignments book. This also demonstrates problem solving & visualization throughout the book. This also demonstrates organization and structure on the first pages, because it has to fit the drawing. I learnt threw steps how to draw 3D.

Overview: ...The spatial geometry notebook helped me alot with understanding, out of anything. We used it to learn step by step to learn how to draw and lable the objects. I learnt how to do spatial geometry by practise everyday, and double checking my answers. I think that my progress in this unit has been pretty good, I did good on the assignments and also good on the tests.

The degree of specificity about Erin's thinking and learning processes differ in each of these examples, but the commonality among them is that Erin is beginning to vocalize what she learned, how she was learning, and what she was thinking. Although the statements are limited because Erin primarily listed what she learned, she is beginning to say how she learned to diagram three-dimensional objects.

These examples of Erin's early attempts at being explicit about her thinking and learning demonstrated a shift in Erin's use of words. Her success was in making limited statements about her thinking in mathematics class. When prompted in our second conversation to show a successful moment, Erin pointed to the question on her fifth test as an example of good thinking.

E: Okay. So, I figured out how much was getting paid. Right? [Mm, hmm.] About how much, over 3. I just made all different packages and then I found out which one was cheapest. All the different ways, like, sets. For, how many did she want?

J: I think 7.

E: Yeah. These are the ways for 7 t-shirts.

J: Okay. So how does that show good thinking?

E: 'Cause I {inaudible words, then a little laugh}

J: Okay. Writing all that stuff down. Do you think the thinking is in deciding between these last few prices? Or is the thinking in doing the calculations?

E: Thinking is in, hmm, probably in the last part.

J: In the last, in comparing the last prices?

E: Yeah. Well, no, that's not thinking. Well, yeah, probably figuring out the packages.

...

J: So, is that thinking when you're just using a calculator?

E: No.

J: No. So, the thinking was in finding the packages and then the rest was just?

E: Calculator.

In the above examples, Erin is demonstrating a beginning metacognitive ability; this ability is nascent—in the act of coming into existence and in the process of being established. Erin's metacognitive statements are not robust, evidenced in the limited nature of her descriptions, but she is in the process of developing metacognitive awareness. Erin's interactive writings and Spatial Geometry Portfolio indicate a movement away from silence because Erin was beginning to recognize her thinking as a successful moment in mathematics class and is communicating her thinking to me.

Erin's success existed on two different levels. First, she exhibited an ability to reason mathematically. In her solution to the test question, Erin systematically considered all the pricing details and used mathematics to support her decision-making. Erin viewed her success as the mathematical thinking she did to complete the solution. She was developing confidence in herself as a mathematical thinker, viewing success in mathematical reasoning as a success in mathematics class. Second, Erin was making initial attempts to engage in metacognition. Schoenfeld (1987) describes three different branches of metacognition: (1) thinking about one's own thinking, (2) self-regulation, and (3) beliefs about the nature of mathematics. Erin demonstrates the first branch by telling me how she thought step-by-step through the test question (systematically creating different packages of t-shirts to compare). She demonstrates the second branch by stating how she learned skills in spatial geometry through practice, double-checking, and a step-by-step process. Erin demonstrates the third branch of metacognition by considering mathematical reasoning as thinking rather than arithmetic computations. By engaging in a metacognitive discourse, Erin was making self-referential statements about her thinking, demonstrating success as a movement away from silence.

Erin's nascent metacognitive ability was a success that was unique to her experiences. Each of the learners in this inquiry experienced particular successes; however, as I considered these successes, I began to notice a commonality among them. The generalized theme of success that I drew from the data illuminated a new voice-stance that was indicative of all the learners' success. The movement away from a stance of silence embodied the *emergence of voice* for each of the learners as the essence of their success in Consumer Mathematics. The final segment of Erin's story of success demonstrates a voice-stance of *emergent voice*, which indicated a movement away from learning in silence.

A SUCCESSFUL STANCE: AN EMERGENT VOICE

In the final portfolio of the semester, students were given control to shape their document to demonstrate their good thinking and learning, and how they had improved over the semester. Erin, who had taken seriously each of the unit portfolios, approached her final portfolio with similar rigor. She selected carefully, often from previous unit portfolios, pieces that demonstrated her good thinking and learning. Consider, again, Erin's final portfolio overview reflection that began this chapter, and the following item reflections from that portfolio:

Item 2: Spatial Geometry Assignments. I included this item in my portfolio because I got stuck in this book a bunch of times on some of the questions, got me thinking. This demonstrated learning math by learning how to rotate and reflect items. One strategy I used in this was to make up little lines showing which way to flip or if it was to be 180 to turn it to the right 2 times, 270 turns three times ect. This shows that I'm a good math learner because I remembered all the rotation and reflection tricks so that I could do a questions without the blocks. It connects with the real world when you need to turn something to a certain degree ect. This demonstrates problem solving and visualization. It also demonstrates organization and structure because you have to remember how to flip or rotate.

Item 4: Comparing Cell Phone Costs. I choose this item because I also did some good thinking in it when I seen it in my portfolio I grabbed it out right away. I learned how to compare a whole bunch of different cell phone bills to see which one was best because all of them have twist and turns to them and they take awhile to compare but it is worth it in the end to get your moneys worth. A strategy that I used in this was to get all the extras off the final cost before you compare them. This shows that I'm a good math thinker because you have to have a certain amount of money without going over and to get a plan from each company with a certain amount of money isn't exactly easy, so that's why I did some good math thinking. You would use this in the real world for example when you are shopping and you go store to store comparing prices to get the best buy. This demonstrates reasoning and problem solving because you have to get the cheapest company for the amount of money that you are given.

The length and substance of Erin's reflections had progressed significantly, even from the Spatial Geometry Portfolio she had constructed halfway through the semester (notice the difference between the reflections for Items 2 in both portfolio excerpts). Erin's final portfolio demonstrated her capstone of success in Consumer Mathematics as she wrote with an emergent voice.

Emergent voice is a type of voice-stance, indicating movement away from silence toward internal, mature voice. Through the process of noticing similarities in the way learners spoke and in what they said about their thinking and learning, I identified three characteristics of *emergent voice.* As Erin and her classmates used words to be successful mathematical thinkers and learners, they came to be: (a) *vocal,* (b) *verbal,* and (c) *intentional.* The students came to be *vocal* as they used words to say things they believed were worth saying. The students came to be *verbal* as they chose specific words to point toward the things they believed were worth saying. And finally, the students came to be *intentional* as they used words to affect themselves and their context. In what follows, I use Erin's final portfolio reflection to illustrate the three characteristics of *emergent voice* and its nascent quality.

Erin was vocal. The first characteristic of emergent voice is that the individual is *vocal.* Being *vocal* means that the students felt that they could speak out, or say things aloud (in writing or orally), and that they did speak out. Because the absence of words was related to a stance of silence, the fact that students, like Erin, were saying things about themselves and their learning became a foundational element of *emergent voice.* Within the portfolio reflection, the element of being *vocal* is demonstrated as Erin wrote about a variety of successful moments she had in Consumer Mathematics. She repeatedly made self-referential statements about how she learned, how her strategies worked, and that she had constructed many of the strategies herself. Speaking out meant that Erin had the confidence to write about her learning strategies.

As Erin's voice emerged, she began to say more things to me about the quality of her learning. For instance, Erin evaluated that she had learned well in the course because she "got the hang of everything pretty good." This statement demonstrates the evolving nature of Erin's positioning with authority. Belenky et al. (1986) noticed that individuals who are in the process of gaining a voice are also beginning to see themselves as their "own authority" (p. 54), rather than relying on external authority to tell them what to believe and to give them knowledge. Erin was becoming an authority on her learning and thinking, rather than agreeing with the success I articulated in our conversation, as a result of her emergent voice. Her authority is also evidenced in her evaluation of learning strategies that were successful and that she would continue to use in future mathematics courses.

Freire (2000) emphasizes the significance of being *vocal* when he states, "Human existence cannot be silent . . . human beings are not built in silence, but in word" (p. 88). To begin to emerge from silence, Erin and her classmates needed to first say something—and the content of what they said was not as significant as the fact that these students were beginning to say things to themselves and to others. As the students began to be *vocal*, they were saying with their words that the teacher was not the sole authority on their thinking and learning. They were becoming aware of their thinking and learning, and through their use of words were developing a sense of authority.

Erin was verbal. Being *verbal* is the second characteristic of emergent voice. Being *verbal* means that the individual is pointing toward specific objects through the selective use of words, rather than just putting words to thoughts and saying them aloud. This process, which Freire (2000) identifies as naming, is comprised of two constitutive elements. The first element is to identify significant objects in the individual's world through a reflective stance. Within my classroom it meant that the students noticed their thinking and/or learning and recognized it as an important cognitive act. The second element is to give a name to the object, which is a label constructed by the individual that points to the object under consideration. Within the context of this inquiry, students were sponsored to write and talk about their thinking and learning, which required them to name their specific cognitive acts.

The two elements of being *verbal* are evident in Erin's words. She used the portfolio reflections, especially later in the semester when she was more *vocal*, to consider the strategies she was using to learn effectively in Consumer Mathematics. As she recognized these strategies, she described them, earlier with the specific steps and then later with labels to identify a certain strategy. One of the learning strategies that Erin named was the idea of "making up little tricks." She used this label to refer to her process of noticing similarities among practice questions and solutions, and then constructing her own way of completing the question. In her Item 2 reflection, Erin makes specific reference to the "rotation and reflection tricks" she used to successfully complete spatial geometry questions. As Erin wrote her final portfolio reflection, she used the name to point to her learning strategy, further refining the use of the name (Searle, 1983).

By speaking out and writing about her successes, Erin had already begun to develop authority in her emergent voice. However, with the additional element of being *verbal,* Erin became an author of her own cognition. An authorial stance required Erin to construct words or phrases, like "making up little tricks," that were meaning-filled for her and that she could use to interact with the world around her. As her classmates also engaged in naming their thinking and learning, they also moved beyond their stance of silence as they began to author their own success. The authoring was at once retrospective as they talked about what they had done well, and prospective as they were beginning to say how they were succeeding in mathematics class.

Erin was intentional. Intentionality is the third characteristic of emergent voice. Being *intentional* means that individuals say things to themselves and to others with specific purposes. These purposes place intentionality behind the things individuals say. The learners in this inquiry developed intentionality as they chose what they said (the content) and how they said it (the words they used). Although some similarities to being vocal and verbal exist, learners were using their nascent abilities of speaking out and naming to say things to a specific audience with the intent of affecting the audience.

The concept of *audience* is central to the intentions of students because the perceived audience affects their intentions. The students sometimes said things to themselves, intending to affect their success, when they viewed *self as audience.* Belenky et al. (1986) observed that individuals who were gaining voice would "engage in self-expression by talking to themselves" (p. 86). Emergent voice needs to say things to self in order for the individual to internalize, author words of significance, and explore intentions imbued in statements to self and others. The students viewed *others as audience* when they said things to someone else, usually to me, about their thinking and learning, intending to affect the teacher-with-learner relationship. In this case, the students often believed they would help me understand them and notice their success in mathematics class. *Emergent voice* needs to say things to others in order to establish the individual's authority, to make intentions explicit to others, and to make self-referential statements.

Erin was writing for both audiences in her final portfolio. Writing about her learning strategies demonstrates *self as audience* because she wanted to positively affect her learning by remembering the learning strategies as she said them to herself. Her intentions were directed at improving her learning, and her emergent voice was *intentional* in supporting her success at mathematical learning. Erin also was intentional in what she said to herself when she was in the act of learning mathematics, evidenced by her strategy of "making tricks" as she was in dialogue with herself with the purpose of learning a new mathematical skill. When Erin wrote that showing all her work was a beneficial strategy because "you let the teacher see that you know what you are doing," there was a shift to *others as audience*. Erin is communicating to me, her teacher, that she not only completes questions methodically to make her own tricks, but also to show me what she understands. I believe that Erin was telling me about how she was successful at thinking mathematically so that I would come to understand her better as a thinker—intending to affect the teacher-with-learner relationship in order to be successful in mathematics class.

Erin's voice was nascent. The three characteristics of emergent voice were indicators that the voice-stance of the students was evolving, from voicelessness to emergent voice. However, being *vocal, verbal,* and *intentional* are not necessarily small steps away from voicelessness, each requiring sophistication in their use. While the idea that learners spoke with an emergent voice was a significant success for them in mathematics class, their voices were just *emerging*. I deliberately constructed the label of *emergent* to convey that the learners did not come to a fully refined voice by the end of our semester together.

Emergent voice is nascent because of its non-autonomous nature. Non-autonomous refers to the limitations of students to say things with their emergent voice on their own, without guidance from another. As I mentioned previously, Erin's thoughtfulness in her final portfolio reflections demonstrated to me the success she experienced in Consumer Mathematics. In several of the reflections she included the statement, "I am a good math learner/thinker because . . . " completing the sentence with metacognitive statements. In our final conversation, I decided to ask her about these statements:

> J: When you think of using your math this semester, did you, did it even cross your mind to think about how you were thinking or how you were learning in the course?
>
> E: No.
>
> J: But, did that become a part of what you did all semester?
>
> E: I guess. Yeah. I guess.
>
> J: Can you say a little bit more about that?
>
> E: I never really thought about that.
>
> J: Yeah?
>
> E: I guess I thought about the marks and asking questions and stuff. I don't know.

J: Mm, hmm. So that idea of thinking about your thinking and thinking about your learning, does that have anything to do with being a math thinker or a math learner?

E: Yeah. I guess. Again, {pause} umm, I don't know. {little laugh}

J: You're not sure?

E: I don't know. I guess it connects with both of them.

J: Yeah. Which one does it connect better with? Does it connect better with a math thinker or does it connect better with a math learner?

E: Ahh. Probably math, {pause} thinker.

J: Okay. How did you, why did you pick that one?

E: I don't know. {little laugh} I just picked that one.

J: Why did it kind of stick out a little more to you?

E: 'Cause, I don't know. It just stuck out more to me.

Erin had written metacognitively frequently in her portfolio reflections, but in the conversation she was unable to notice these cognitive processes. I believe that if Erin could not notice the processes that she had already engaged in, being metacognitive would also not occur autonomously.

As disappointed as I was in Erin's response to my prompts about her thinking about metacognition, it demonstrated to me that emergent voice is non-autonomous. Erin could engage in metacognitive thought when the prompts were within her ability to respond to, but could not distinguish this cognitive act from others to engage in it autonomously. Erin's response caused me to wonder why emergent voice is non-autonomous. Vygotsky's (1978) concept of zone of proximal development was effective in framing the non-autonomous nature of emergent voice. The emergent voice that I was noticing in Erin and her classmates was within their zone of proximal development, meaning they could engage in refining and using their emergent voice within the guidance of their teacher. Erin brings into view a process that was "currently in a state of formation, that [was] just beginning to mature and develop" (Vygotsky, 1978, p. 87). It might seem like a step back for learners to be non-autonomous in their use of emergent voice, but the nascent quality of emergent voice almost necessitates an incompleteness and is what distinguishes emergent voice from mature and internal voice. In other words, emergent voice does not experience consolidation because of its *nascent* quality. What Erin demonstrates is that emergent voice is vulnerable and continually needs to be fostered by another who recognizes the importance of voice and who has voice already—their teacher.

Emergent voice can be identified through the three elements that characterize the speaker's use of words in that voice—being *vocal*, *verbal*, and *intentional*. Although in Erin's example all three characteristics of emergent voice were present simultaneously, this is not a requirement of emergent voice. Rather, at least one characteristic must be present to indicate a student's movement away from voicelessness toward emergent voice. In fact, some of Erin's classmates experienced different intensities of the three elements of emergent voice, relying on one characteristic over the others. Because this was the students' initial movement away from voicelessness, each student experienced some tentativeness, indicated through the non-autonomous nature of emergent voice, when they spoke. Each of the students in this inquiry experienced some movement away from voicelessness, and the meaning of their success comes from the new voice-stance they formed—speaking with an emergent voice.

THE EMERGENCE OF VOICE AS SUCCESS

It is important to understand that *emergent voice* is not a stage, but rather a continuum along which each student moves, as he or she continually refines his or her sense of success. Emergent voice is recursively both an end and a means, a dynamic process where emergent voice supports and fosters the emergence of voice. The concept of *emergence of voice* is central to the success of the students in this inquiry and was a way of being in our classroom—it was a process of *becoming*. Although the essence of success was emergent voice, the evolution is understood as the process of that emerging. The movement away from voicelessness and engaging in the emergence of voice is not a paradigmatic shift of success and of voice stance. Instead, the emergence of voice was so quiet that it required me to be vigilant in inviting the students to engage in the process of emergence and was only heard as I listened intently to them. As I listened intently, was in conversation with the students, and was thoughtful about those interactions, the students and I negotiated a sense of success that was particular to their experiences and generalized to interpret all of the students' successes.

The *emergence of voice* captures the initial utterances as individuals begin to say things to themselves and to others. It is akin to seeing the first flickers of light from stars in the night sky. Those initial pieces of light seem to be brilliant because of the background of the dark sky. So too, the first attempts of the students saying things, pointing to objects with words, and being purposeful in what they say are brilliant moments against their voicelessness. The first lights in the sky also seem to flicker before they shine steadily, just as the first attempts of students with their emergent voice are tentative. In considering Erin's story of success, the *emergence of voice* can be seen in three strands that demonstrate Erin's *becoming* in Consumer Mathematics.

First, Erin's value of her ideas evolved over the semester from not perceiving value in her own ideas with the fractal cards, to identifying her mathematical reasoning as a success in mathematics, to perceiving her "making tricks" as a successful learning strategy. With the emergence of Erin's voice, she was retaining her original tentative successes and building more complex ones. We could view her successes as concentric circles, where closer to the inner circles was the successful moment where Erin wrote about her thinking on the test question in her eighth interactive writing. As we move a circle toward the outside, the next successful moment contained the first but was more complex because she named her cognition as "making tricks" rather than merely vocalizing that she had ideas of value in mathematics class. The emergence of voice is characterized by the consolidation of previous successes while building on more complex success.

Second, Erin's ability to make self-referential statements evolved from difficulty in voicing her frustration with the fractal cards, to making limited statements about her thinking in the eighth interactive writing, to a final portfolio entry that contains many statements about her thinking and learning. Erin's ability to say things about herself demonstrates that she had moved away from a stance of voicelessness, where the authority figure defines her identity, and had begun to author her own identity as a successful mathematical thinker and learner. By noticing and saying her successful learning strategies, she was actively shaping the way in which she defined herself in mathematics class with her words and in saying words about herself to others.

Third, Erin was in the process of becoming an authority on her learning. By requesting that I tell her about the pattern in the fractal cards, she placed herself subordinate to the authority in the classroom. She moved to describing, in a limited manner, her success on the test question, but ultimately became the authority on her thinking and learning when she provided rationales in her final portfolio as to why she was a good mathematical thinker and learner. These three strands demonstrate the process of the emergence of voice, best understood with the aid of Chickering and Reisser's (1993) metaphor of vectoring to describe the direction and rate of growth. Erin's vector began at voicelessness and was pointing toward a mature voice, while her rate varied in the tentativeness as her voice emerged. Erin's success was the emergence of her voice.

GIVING VOICE TO THE INQUIRER:
MY LEARNING JOURNEY AND *BECOMING*

Through my experiences in this inquiry, I have come to understand that voice is brought forward in interactions with others and emerges within an ethic of care that is "rooted in receptivity, relatedness, and responsiveness" (Noddings, 1984, p. 2). These three characteristics of an ethic of care can be actualized by listening to learners in the classroom. As I drew on my knowledge of each student, I was able to listen to each individual as a student, a learner, and a human being. This authentic listening, enacted through classroom interactions, interactive writing, portfolios, and conversations, provided an opportunity for me to hear both the silence and voice of each learner in what he or she said to me. It was this act of listening that not only allowed me to hear the success of the learners, but also to foster the learners' success. Listening intently to the quietest of utterances from an emergent voice encouraged the development of that voice.

Within this inquiry, I was at once a teacher, inquirer, and learner. As a teacher, I have confirmed the importance of listening to the success of learners. More importantly, I have learned to hear the learners more effectively. I have come to understand the complexity in my classroom, of the learners, our interactions, and myself. As an inquirer, I have learned what it means to systematically inquire into my practice and student learning in the classroom. Although I teach within an inquiring stance, it is significant to me that as I shaped and implemented this inquiry, I learned much about what it means to make meaning of the words of learners. As a learner, I have been engaged continuously throughout this process in learning about the learners in my classroom, about success in mathematics class, about myself, and about systematic inquiry. I found that as a teacher, inquirer, and learner, my quest was complex. The complexity came from the interactions with the learners as I listened intently to them and from the multiple layers of meaning and learning that were occurring in the inquiry and in the classroom. In a sense, the inquiry was part of my *becoming*.

I believe that stories such as Erin's are worth listening to, are valuable to understand, and are worth telling to a broader audience. The stories of students are theirs to be said and theirs to be amplified—it is their *becoming*. Central to narrative inquiry are stories that inform and draw us into understanding the experiences of learners so that we, as teachers, can inform and transform our practice. Listening to learners is vital to their becoming in a mathematics classroom. I hope that others are not only encouraged to engage in inquiring into the success of learners in mathematics classrooms, but that they are also encouraged to listen to the learners that they interact with daily. If we want to understand how to teach exceptionally, to understand how to design curriculum effectively, and to understand what it means to learn excellently, we will have to listen to the voices of the students in our classrooms.

NOTES

1. Throughout the chapter, Erin's authorship has been maintained in its original form, including spelling and grammatical errors.
2. In this chapter italics distinguishes words that I have used to represent conceptual ideas that I identified as a result of my research.

REFERENCES

Baxter Magolda, M. B. (1992). *Knowing and reasoning in college: Gender-related patterns in students' intellectual development*. San Francisco: Jossey-Bass.

Belenky, M. F., Clinchy, B. M., Goldberger, N. R., & Tarule, J. M. (1986). *Women's ways of knowing: The development of self, voice, and mind*. New York: Basic Books.

Chickering, A. W., & Reisser, L. (1993). *Education and identity* (2nd ed.). San Francisco: Jossey-Bass.

Clandinin, D. J., & Connelly, F. M. (2000). *Narrative inquiry: Experience and story in qualitative research*. San Francisco: Jossey-Bass.

Freire, P. (2000). *Pedagogy of the oppressed* (30th anniversary ed.). New York: Continuum.

Manitoba Education, Training and Youth. (2002). *Senior 2 consumer mathematics: A foundation for implementation*. Winnipeg, Manitoba: Manitoba Education, Training and Youth.

Mason, R. T., & McFeetors, P. J. (2002). *Interactive writing in mathematics class: Getting started*. Mathematics Teacher, 95(7), 532-536.

National Council of Teachers of Mathematics. (2000). *Principles and standards for school mathematics*. Reston, VA.: Author.

Noddings, N. (1984). *Caring: A feminine approach to ethics and moral education*. Berkeley: University of California Press.

Schoenfeld, A. H. (1987). What's all the fuss about metacognition? In A. H. Schoenfeld (Ed.), *Cognitive science and mathematics education* (pp. 189-215). Hillsdale, NJ: Lawrence Erlbaum Associates.

Searle, J. R. (1983). *Intentionality: An essay in the philosophy of mind*. Cambridge, MA: Cambridge University Press.

Simmt, E., & Davis, B. (1998). *Fractal cards: A space for exploration in geometry and discrete mathematics*. Mathematics Teacher, 91(2), 102-108.

Van Manen, M. (1986). *The tone of teaching*. Richmond Hill, Ontario: Scholastic.

Vygotsky, L. S. (1978). *Mind in society: The development of higher psychological processes*. Cambridge, MA.: Harvard University Press.

CHAPTER 10

EXPLORING CULTURE AND PEDAGOGY IN MATHEMATICS CLASS THROUGH STUDENT INTERVIEWS

Jesse Solomon
Boston Teacher Residency, Boston Public Schools

I have been a mathematics teacher in urban schools for the last 10 years. I love to teach and am generally considered a successful teacher by others. The more I teach and the more I learn about teaching, however, the more aware I become of the tensions inherent in trying to implement what I call a "constructivist" pedagogy and curriculum. Like teachers everywhere, I search for good matches between pedagogical approaches and the ways my students learn best. As a White man whose students are mostly children of color, it is my sense that many of the issues I grapple with can be better understood through explicit attention to culture. This chapter reflects my attempts to begin to study some of these issues more formally, to move them from being nagging hunches that I think about throughout the day—driving home or laying awake at night—to more precise questions

Teachers Engaged in Research
Inquiry Into Mathematics Classrooms, Grades 9–12, pages 175–196

that I might research. In investigating the interactions of culture and pedagogy in one mathematics classroom, I hope to raise broader questions about equity in mathematics education for teachers and researchers. This chapter does not conclude with a neatly packaged set of answers, but rather attempts to articulate an exploration into assumptions and practices that may contribute to the understanding of complex classroom dynamics. In what follows, I describe the context for this study, explore the origins of my research questions, introduce four students whom I interviewed, and discuss and interpret the students' comments.

BACKGROUND

The School

This study took place in a public charter high school in Boston where I had taught for seven years. The school enrolled roughly 200 students, grades 9–12. Seventy-two percent of the students were classified as African American, 15% White, 9% Latino, and 4% Asian American. The mathematics department used the Interactive Mathematics Project™ (IMP) curriculum for all classes. This National Science Foundation-funded curriculum was produced in response to the National Council of Teachers of Mathematics' *Curriculum and Evaluation Standards for School Mathematics* (1989). IMP is integrated across the four high school years: students study aspects of algebra, geometry, trigonometry and probability and statistics each year. The problem-based curriculum is organized into a relatively few units which are generally focused on a large central question. The mathematics department had been using IMP for the five years prior to this study as part of the school's effort to produce responsible, resourceful and respectful democratic citizens; we sought a curriculum which would ask students to think deeply about mathematics and to explore ideas in a variety of contexts. The school did not employ tracking; all students at the same grade level took the same mathematics course.

According to both internal and external measures, our students achieved reasonable success in mathematics. For example, over 95% of the school's graduates were accepted to college, and 90% of the Class of 2003 passed the mathematics portion of Massachusetts' graduation exam, a rate equal to or higher than all the other public non-selective high schools in the city. However, there was widespread agreement among the faculty that not all of the students were being educated to high levels and not all graduates were prepared for success after high school. Students' SAT scores were low, and some graduates reported a mismatch between their high school mathematics experience and the courses they took in college. Students, parents, and even other faculty members regularly questioned whether the mathematics department's approach was appropriate for the school's student body.

These challenges presented the mathematics department with a set of questions about how to teach our subject. Were we failing to make the adjustments necessary to serve our students well? Should we stick with the reform curriculum and approaches, or return to more traditional approaches? Hiebert (1999) writes, "Presuming that traditional approaches have proven to be successful is ignoring the largest database we have. The evidence indicates that the traditional curriculum and instructional methods in the United States are not serving our students well" (p. 13). We did not feel like we were abandoning a mathematics education system that was working well for our students. We knew we had to explore the extent to which the gaps between our students' real and intended learning outcomes were a product of the school's curricular and instructional strategies, or whether those strategies actually helped reduce the gaps. This chapter focuses on my efforts to investigate these questions in the context of my classroom.

MY CLASSROOM

A typical sequence of activities in my classroom begins with a problem posed by me or from the text, followed by time for students to think about the problem themselves, time for students to talk about the problem in small groups or across small groups, and presentations to the class by one or more students, which generally turn into a broader group discussion. These discussions may end in a general consensus about a conclusion or set of conclusions, a request for a final explanation or demonstration by me, one or more questions that we agree we need to pursue, or some combination of these options. Often the sequence does not coincide with a fixed class period. Wherever we are at the end of the period, I try to make sure we do some kind of summary and reach some closure. The next class period generally begins at whatever point in the sequence we ended the day before.

Ernest (1996) describes the "underlying metaphor" of social constructivism as "that of persons in conversation" (p. 342). I work to establish a classroom culture in which a mathematics course—a year of math classes—feels like a series of ongoing conversations about mathematics. That is, my students and I regularly engage and struggle with ideas, talk them through, present and defend our ideas to each other, question each other, and try to reach some conclusions together. I emphasize that students should talk to each other, rather than the discussion following a teacher-student-teacher-student-teacher-student interaction pattern. My job is to have a clear idea of where we are going, to help us make decisions about what would get us there, and when we seem to be straying, to assess whether the new direction is likely to be fruitful—now or at some later point in the students' mathematical careers. I take great care in making decisions about when to explain or demonstrate a mathematical idea explicitly and when to ask students to work through the material. My approach is based on the belief that students learn mathematics in a more lasting manner and at a deeper level if they reason through it themselves. I refer to what I have described in this section as a "constructivist" learning environment.

Rethinking My Approach

A few years before I began this research my teaching style was, by my own account, furthest on the spectrum toward being solely a facilitator. I had worked hard to develop a classroom where students did much of the mathematical heavy lifting. Observers would comment that they were impressed with the students—their thinking, their ability to express themselves, their engagement with the mathematics. Yet the students were asking me to inject more of myself and my mathematical thinking into the conversation. Freire's writings had been extremely influential in my formation as a teacher and in my thinking about pedagogy, and re-reading his writings made me rethink my approach:

> When teachers call themselves facilitators and not teachers, they become involved in a distortion of reality. To begin with, in de-emphasizing the teacher's power by claiming to be a facilitator, one is being less than truthful to the extent that the teacher turned facilitator maintains the power institutionally created in the position. That is, while facilitators may veil their power, at any moment they can exercise power as they wish. (Freire & Macedo, 1995, p. 378)

This led me to ask myself if I was engaging in a "distortion of reality." Was my teaching style, my attempt to help students become mathematically powerful, actually a veiling of my own power? Adding to this, Walker (1992) writes that a facilitative teaching style "models the learning style of the mainstream. It is oppositional to and noninclusive of the cultural norms of teaching that have been documented as being valued by some African-American children" (p. 323). In fact, Foster (as cited in Walker, 1992, p. 323) reports that some African-American students view teachers who employ this style as engaging in a "privatization of knowledge." I knew there were times when I was not explaining, summarizing, or correcting, when doing so may have been extremely helpful and even necessary to some or all of the students. I wanted to get better at recognizing the times when the class would benefit from my being more direct, without making this type of instruction the dominant feature of the class.

Another impetus for rethinking my approach occurred when I observed a debate in one of our school's history classes in which certain students performed skillfully and others floundered. It struck me that many of the differences in performance may not primarily be a function of the students' immediate preparation. Rather, there are a complex set of "moves" that go into making a strong argument, and students come to class with varying degrees of familiarity with these moves. Ochs, Taylor, Rudolph, and Smith (1992) posit that "there is at least the potential that children sitting around a dinner table listening to and collaborating in ... storytelling theory-building [that is, theory-building through storytelling] are being socialized into the rudiments of scholarly discourse" (p. 68). Given that scholarly discourse such as debate gets practiced regularly in certain households and families, in what ways does the ability to participate in such a discourse affect students and how they are perceived in school? Delpit (1995) writes, "I found that the people who appear to be discovering everything on their own have actually received direct instruction at home ... All day long there is direct instruction that middle-class parents provide for their kids." Since I was raised in what I might call a constructivist household—one in which I was encouraged to build and develop my own understandings and ideas—I wondered what set of skills and habits I had learned at my family's dinner table. Might I be assuming that students in my classes were bringing some of these same skills? Was I failing to grasp the need to teach certain skills because they were so much a part of me?

I wondered whether there are analogous "debate moves" in mathematics. Could I help students acquire power more readily if there were certain pieces I just told them? Moses and colleagues (Moses, Kamii, Swap, & Howard, 1989; Moses & Cobb, 2001) eloquently make the argument that mathematical competency serves as a gateway to power in our society. Delpit (1988) claims, "If you are not already a participant in the culture of power, being told explicitly the rules of that culture makes acquiring power easier." Delpit (1986) also comments that:

> Writing process advocates often give the impression that they view the direct teaching of skills to be restrictive to the writing process at best, and at worst, politically repressive to students already oppressed by a racist educational system. Black teachers, on the other hand, see the teaching of skills to be essential to their students' survival... [Their] insistence on skills is not a negation of their students' intellect ... but an acknowledgment of it. (p. 383)

How was this dynamic playing out in my classroom? In an attempt to facilitate rather than restrict students' thinking, was I leaving out essential skills? To "tell" students explicitly, however, presents another dilemma. Delpit (1992) wonders whether White teachers

> question whether they are acting as agents of oppression by insisting that students who are not already a part of the "mainstream" learn that discourse. Does it not smack of racism or classism to demand that these students put aside the language of their homes and communities and adopt a discourse that is not only alien but has often been instrumental in furthering their oppression? (p. 297)

Delpit warns that this question—*How do I teach my students what they need to know if I conceive of the act as oppressive in itself?*—can be paralyzing. I was struggling to resolve this dilemma in a manner that allowed me to teach important mathematics in a way that was accessible to all my students.

Research Questions

My interest in investigating these ideas in the context of my own classroom led me to formulate the following questions:

1. In what ways do two "leanings"—which I will label *constructivism* and *explicitness*—in my teaching affect student learning?
2. What are the cross- and inter-effects of these two "leanings" (i.e., on each other and on the student as a result of the combination)?
3. In what ways does a student's culture(s) affect his or her "mesh" with these "leanings"?
4. In what ways does my own culture illuminate for/blind me to these effects?

I used these questions not as strict boundaries for my research, but as a starting point to begin a broad investigation. The questions arose from a combination of trends I had noticed in my teaching and the writings of various theorists and researchers. The following describes how I delved into this research and some of the results I found.

The Participants

As a way to begin to address some of these questions, I interviewed four of my students, all of whom were seniors. I chose students I had come to know as "informants," students who had shown a comfort level talking to me about issues surrounding school and our class. All four of the students were passing the class at the time of the interviews, although none were excelling.

TIA

Tia described herself as an "average, Black female." She is a delightful young woman, cheery and earnest, who hails from a religious family. When I asked her about the things she most values, she said:

> Family that sticks together. I know people who don't have any family, just a mom and dad, they don't have no cousins, no uncles. For me, I just want to be around my sisters, cousins, uncles, aunts, cousins, family, family, getting to know each other. We all live in the same community. Family, quality time. How am I going to have a cousin over there and not know him? You got to bond. Have those big family reunions, cookouts. Family, I love family.

Tia's family moved from Florida to Boston while she was in middle school. Tia commented on some of the differences she experienced:

> Schooling is much different than the schooling down there. When I got here, and I went to school, I was like, "Where are all the White people?" Whereas down there, ... it was something like on an equal level. ... In most of my classes up here, I only have like two or three White kids in my class. ... I liked the equal setting better. ... Maybe [the school system isn't] trying to segregate us, but it is pretty segregated. And I don't think it should be like that, because we fought against that.

Tia was an active member of class, regularly asking questions and throwing out ideas. She made it clear that she wanted to be successful in the course. I viewed her as a strong member of small groups, able to work well with most of the students in the class.

TANISHA

Tanisha is very much her own person. She stated, "I want to have my own style. ... I want there to be something different about me." When I asked her how she defines herself culturally, she said:

I don't think I have a culture. ... If I was to say I'm African-American, like there's a lot of Africans or whatever they, they would do their culture where like I really don't know about the culture. ... I don't have a specific culture, ... I say I'm American-ized, which means we get to do anything we want really. ... I wish I did have a culture, I want to be in a culture, but I don't have one. Because I'm not around,— I mean, I haven't had enough time to be around a culture.

She described African-American culture as being "so varied" and hard to define as any one set of things. She commented, "I think I have an attitude, like a strong attitude. ... We're [her sisters] not gonna let nobody stop us, like we're persistent, whatever we say, we're just gonna keep going by it."

In class, she spoke very fast; other students often had to ask her to slow down, or to repeat what she had said. She often would say something minimal initially and then expand only when others asked her to do so. She was at times very involved and the class would feel almost centered around her. Other days, it seemed that her mind was elsewhere. I saw her as a student who told it like it was. She had a strong sense of how she was affected by her race. For example, she told me about how in her past schools, she could just "sit and look nice while the other people were playing around," and that the Massachusetts' high-stakes exam was about race, "Because how many minorities live in poverty, and how well are they getting educated?"

JENNIFER

Jennifer attended a parochial middle school, and so held a slightly different expectation about school and education than many of her peers. She was generally studious and wanted to do well. In class, she was usually polite and fairly quiet. She would talk with friends, sometimes devoting more attention to her friends than the mathematics. Outside of class, or if we had relative privacy in class, she would tell me what she was thinking, whether she thought I wanted to hear it or not. She took it upon herself to give me advice from time to time about how to make class better—"Mr. S., you should do more things like we did today"—an act I found particularly helpful. Jennifer hailed from Haiti. She told me that she thinks many Haitian families come to the U.S. to get a better education.

She described her home as preserving Haitian culture: her family speaks Creole and usually eats Haitian food, the house is decorated with paintings from Haiti and the Haitian flag. Jennifer talked about taking pride in being from one of the first Black countries to gain its independence. Her family attends two different churches. Her father attends a traditional Haitian, United Methodist church, which Jennifer describes as extremely conservative; the service takes place in French and "everyone just sits there." Her older sister attends another church that Jennifer describes as being "freer"; the services are in English, she can understand the pastor, gospel music plays a large role, and people "sing and dance and shout." Her father describes this church as a "party church." Jennifer works at a shoe store after school, where she reported that she was always asked to wait on the "foreign people," no matter where they are from.

GREG

Greg was a fairly quiet student, whose face often did not betray what he was thinking. At times, I would come over to him in class because I thought he was distracted, but he would be thinking about the work and had a question. He did not appear to be actively pursuing the information—by asking other students, for example—but genuinely wanted to learn and would listen intently when I or another teacher or peer worked with him. He told me that he had always been quiet, but that he had come a long way. He described growing up as being characterized by an almost painful shyness, and credits an early teacher and an experience at a summer basketball camp with helping him break out of his shell. He describes himself as African-American. When I asked him to describe his culture and the traditions or habits that feel important to him, he said, "It's hard to put into words. ... I never been asked this question." Finally, he told me, "You were taught a certain way, then when you get into the real world, the real life society, then you see why you were taught that way." He told me that he was taught "to always stay true to yourself. To not let society change you, where it's gonna lead you to the wrong road." Greg's finishing high school was very important to his mother who ran an extremely strict home. Greg had four siblings and lived in what he described as a crowded house of six. He reported that each child had "a separate time with our mom."

METHODOLOGY

This research is based in a tradition of practitioner-research, which systematically examines issues of equity through careful analysis of daily dilemmas and phenomena (Ballenger, 1998; Gallas, 1994). Researchers using this tradition attempt to think and learn about race and culture without allowing generalizations to overtake the unique nature of the subjects or the classroom setting. The interviews were informal, each lasting roughly an hour. I asked the students a set of questions about good learning experiences (inside and outside of mathematics class and outside of school), culture and race and how they played out in school, when they felt most involved in mathematics class, and their best teachers. I then went through an iterative process of reviewing the tapes, sharing and discussing the data with colleagues in a teacher-research group and with other researchers, and then returning to the tapes and the students' words for further examination.[1] In what follows, I describe and illustrate three themes that emerged from the interviews.

THE NATURE OF MATHEMATICS

While I did not explicitly ask students to describe the nature of mathematics, many of their comments painted a coherent picture of their notion of the field. In the following, I discuss some of the common themes that emerged during the student interviews.

Mathematics as a Step-by-Step Endeavor

The comments of two of the students in particular, Tia and Tanisha, described the study of mathematics as learning a set of steps. Tia described how she learns: "I have to like see what it is saying step-by-step, and then, if you put it before me, then I may be able to do it more successfully than you telling me, writing it down." Tanisha described a previous mathematics teacher whom she considered effective: "So he writes it on the board. And then he goes by step-by-step. Like he'll write step one, that's step two. You're not going to skip a step. He goes through each one with you."

Both students used the phrase "step-by-step" to describe the presentation of the mathematics. Tia commented that she is more successful if she is able to see it "before" her, while Tanisha's teacher was effective because he ensured that "You're not going to skip a step." They describe a field made up of a set of algorithms that require the successful execution of sequential steps; if one follows the steps faithfully, it will lead to the right answer. Boaler (2002) describes many students in traditional beginning algebra classes making similar comments, for example:

> He teaches you exactly how to do it. He'll show us tricks that he knows. And he'll always give us the easiest way ... You don't have to have big thoughts. There's simple little steps that you just follow. ... Make it, like a habit to do it. Just don't even think about it. Just like, okay, now what are the next steps? (p. 8)

Ideas about Mathematics Curriculum and Instruction

The students said that the mathematics program was both not what they were used to and not what they expected. They expressed anxiety that the IMP materials were "experimental," and as such might lessen their chances of success in certain key areas like standardized tests. Jennifer explained:

> It's not like basic math, I don't know. I'm used to, like, seeing more numbers, like in elementary school, there's numbers in the math book. And then in middle school, numbers in the math book. [Our school] ... problem solving, a whole bunch of words and POW's [Problems-of-the-Week]. ... It's just like it makes everything so difficult.

The implication is that mathematics is supposed to be a set of problems consisting largely of numbers and symbols. The IMP curriculum had plenty of numbers, but a typical textbook page consisted mostly of written text rather than a set of problems. The questions were embedded in the text, and there were fewer of them.

Jennifer described the reaction of two trusted people—her father and a friend—to the curriculum:

> I'm doing my homework and my Dad's like ... "How you gonna do this stuff? I'm used to math with numbers." So then we have to like sit there and both try to understand it. When he says stuff like that, that's when I'm like, "It's really true that we need math with numbers." ... When I would do homework, like my friend would come over she'd be like, "Jennifer, what are you doing?" ... And I would see her math stuff and my math stuff, and I'm like, "There's no way this stuff is going to help me on the SAT better than any other type of math program."

Jennifer's father and friend gave her the feedback that what she was doing was different than the norm. Jennifer's family immigrated to the U.S. in large part to gain a better education for the children. Jennifer did not feel that she could reassure her father about the relevance of the mathematics even if she wanted to; her friend's questioning reinforced her notion that there are other schools nearby actually doing the mathematics she would need.

Tanisha echoed this feeling in a story about a visit to another school. At the time of our interview, we had been working on a curriculum unit that develops trigonometric ideas through the examination of the motion of a Ferris wheel.

> Like the sine stuff now, I seen at another school when I went to SAT (Prep class), but they basically simplified a traditional way, and then I'm like I understand it, it is related to it, but I like their way, still, cause I understand how it's related to this math now.

It took seeing the material in another setting for Tanisha to understand that the mathematics we did in our class was "related" to more traditionally-taught mathematics, although she would still rather encounter it in its stripped-down form. Tanisha said that this experience made her realize that "we're not doing nothing for nothing, cause that's how I used to feel. Doing nothing for nothing, talking about cubes and cookies. I understand it, how it is related."

Tanisha's phrase, "doing nothing for nothing," and the deep emotion behind it is striking. Tanisha wanted to learn mathematics so that she could be successful, yet she came to class every day to engage in a pursuit that she did not ultimately think would be relevant to her goals. Tanisha was not questioning *how* the study of mathematics would ever help her, but rather how *this* approach to the study of mathematics would help her. Given her view of mathematics as instrumental to her future success, she resented studying it in a context for which the purpose was unclear. Similarly, Jennifer added:

> The big question is: how is this going to help me in the future and stuff? And we already feel like half the stuff we do in math isn't going to help us in the future, especially if it is not like something you plan on pursuing.

It is important to note that both of these students expressed an understanding of the rationale behind this approach to mathematics. Jennifer said that the mathematics "makes me think more, like when I see a problem," and Tanisha said, "That's what they are trying to do, make me learn." While they understand the aims of this approach, they do not seem to equate "thinking more" with what it means to do mathematics, or at least useful mathematics.

Mathematics Embedded in Language

As mentioned above, our textbook was text-rich; the problems were embedded in stories and written descriptions. Tanisha remarked:

I never liked word problems, so the four years I've been here I have to do mostly word problems. And sometimes, like when I see the reading, I'm just like, "I don't understand this, so I'm not going to do it." That's what makes me just don't do it, like if I don't know what they're talking about. So I don't want to do it, because there's an easier way or something.

Jennifer characterized the curriculum as "a whole bunch of words and POW's and stuff like that. It's just like it makes everything so difficult." She said that it is hard because "You have to try to relate it back to the way it was taught to you, and like all the wording and everything like that, because we do it all by like definitions." Both students described the struggle of decoding the words and extracting the mathematics. While the curriculum was designed to be more integrated and draw more upon the "real world," it also seemed to throw up additional barriers for students, especially for students with weaker language skills. My sense was that the students often felt a need to translate a problem out of the curricular context and into a format with which they felt comfortable, which usually was a stripped-away algorithm. Only after solving the problem using the algorithm would they (sometimes) translate the answer back into the curricular context, as if they understood that providing the answer in context was a necessary piece of success in the curriculum, but not necessarily related to understanding mathematics.

What Should We Make of these Students' Comments?

The students had a view of the mathematics that we were doing as separate from the mainstream. Jennifer's father asked, "How you gonna do this stuff? I'm used to math with numbers." She thought, "There's no way this stuff is going to help me on the SAT better than any other type of math program." Tanisha thought of it as "doing nothing for nothing, talking about cubes and cookies." These students raised questions about the role of mathematics in terms of power in a society. They had perceptions of the "real world" and the ways in which mathematics could be useful, and they wanted their mathematics program to align with or inform that vision. Mathematics, as described by these students, is not at its core a collection of powerful and elegant ways of thinking that can help one puzzle through all sorts of difficult and compelling problems. The students seemed to see that such ways of thinking might be one possible outcome of studying mathematics, but they did not frame such thinking as the ultimate goal. One implication of such a view of mathematics is that the teacher's goal of developing students' ability to construct meaning and arguments competes for these students' attention with more immediate concerns such as college entrance exams. The issue is not that these students do not want to think or work hard to succeed—they do. Rather, they have certain conceptions of what it will take to succeed that contrast with the messages they receive from the curriculum.

Taken against the backdrop of Delpit's writings, these comments raise further questions about these students' and their families' relationships with school. I often felt in my conversations with families that they expected and even trusted our school to teach their children some of the rules of mainstream culture and power. The students' comments have led me to examine whether there were dynamics in my classroom that were perhaps not a betrayal of that trust, but at least a failure to meet the expectation.

VIEWS OF TEACHING AND TEACHERS

A second set of questions I asked the students focused on their perceptions of effective teachers and teaching, both in our mathematics class and in other classes. Below I discuss a typical teaching "move" of mine and the students' preferences about seeking and receiving a teacher's feedback.

Turning Students' Questions Back on Them

I often respond to students' questions that I believe are premature (i.e., that a little more thinking on the student's part would yield progress) by asking them to continue working on the problem, rather than giving them a direct answer. Greg and Tanisha commented on this move in their interviews:

Sometimes it's aggravating, because ... that's the reason why I'm asking the question because I don't understand it. So, um, that's aggravating sometimes, but like, you gotta live with it.

Sometimes it's very agitating, but not like, like if we was really knowing, but if we really, really don't know it, like none of us did the homework, or we don't know what it's talking about, then it's really aggravating, 'cause neither one of us is really trying our hardest to understand ... why he won't tell us the answer.

I regularly have conversations both with the class and with individual students in which I try to be explicit about the rationale for this move and explain how it stems from my belief that they are capable of figuring out the answers. The students, however, often view my deflection of their questions as indicative of my not helping or worse, not caring. Knowing that they may think this way only complicates the issue. I do believe that part of teaching is to put students in uncomfortable intellectual situations and then support them as they resolve conflicts and gain new understandings. Yet my explicit denial of short-term help, which I could easily give, has at least the potential to harm our relationship.

Direct Feedback

When I asked Greg how he learns best in mathematics class, he said he most appreciates when a teacher comes over to him and has a one-on-one conversation. "I can tell you (the teacher) ... everything I'm thinking at the time and like show you ... what I'm thinking. Then you can tell me what I'm doing wrong, what I'm doing right." When asked to think of one of her best learning experiences, Tia talked about preparing for a history presentation:

> I could do a pre-presentation. ..., And then she [the teacher] would ... look it over and give me feedback. And then it was like I had it done, it was good, I had it down pat. And then when I got up there, and it was really time for the real (presentation), I did well. And I thought that was a good thing for me because I didn't always have to look down at my paper because I knew what I was doing and I knew what I was presenting.

These students described a preference for a private time in which to try out ideas, receive direct feedback, and build confidence before going public.

What Should We Make of These Students' Comments?

I hear in my students' requests not a plea for things to be made easy or to avoid working hard, but rather a description of how they think they will learn best. I know that I have a number of habits in my teaching—such as turning a question back on a student, or telling a student to work with his group before calling me over for help—which at this point are almost automatic. Much of my teaching is based in a desire to convey constantly the message that the students are fully capable, and yet I may be doing things that are inconsistent with a particular student's knowledge about how he or she learns best. Such communication gaps may have resulted from practices and approaches I regularly discussed with students, but that they may not have accepted (perhaps this is the pedagogical equivalent of "I taught it; they just didn't get it"). I meant to have a hard line about the structure of class; I meant to embed in everything we did the fact that the students were smart and would be expected to act like it every day. These interviews have called into question whether I should take such a hard line. In being so doctrinaire I may have overlooked chances to help kids learn.

Students seemed to be asking for a kind of hybrid mathematics class. There were times when they wanted to be told, and others when they wanted to puzzle things through for themselves. While this may seem like common sense, it requires a flexibility that runs counter to the structure of the curricula. The curriculum we used was so well thought-out that it was all but scripted. The accompanying message is that teachers should teach the curriculum just the way it was written. The developers urged us to trust the materials, ensuring that it would all fit together. While developers can control whether the concepts fit together on paper, this is less true when it comes to ideas fitting together in students' minds. The curriculum included sections that attended to students' differing learning styles; however, the curriculum did not seem to take into account the notion that the whole approach, even with various accommodations, may require significant alterations to reach all students.

The interviews helped me to rethink the place of student thinking in my classes. I had been extremely vigilant about my own direct contributions to the class conversation, fearing that I might be limiting opportunities for students to think on their own. I am now more likely to insert my own thinking earlier in a class discussion, while remaining aware of how such an action affects student thinking. As I try to create an environment in which the students are doing much of the work, I try to use my mathematical insertions to explain or clarify, or to give students something with which to wrestle, without cutting off their thinking or encouraging them to try to mimic me.

It is interesting to examine the themes discussed so far—the nature of mathematics and preferred teaching styles—together. If students see the field of mathematics as a fairly fixed entity without much room for one's own thoughts, a teacher's facilitation (to make a simplified contrast with direct instruction) may be even more frustrating. Freire and Macedo's (1995) comment about facilitation raises the question: Does a student in this situation interpret the teacher's actions as disingenuous? The student understands mathematics to be a set of processes. The teacher obviously knows these processes but refuses to divulge them. No matter how much a teacher describes mathematics as a complex field to be wrestled with and understood, the student who does not believe that notion may experience certain classroom interactions as disingenuous.

STUDENTS' OWN LEARNING STYLES AND EXPERIENCES

In order to try to learn more about pedagogies and learning situations that the students experienced as successful, I asked them about other institutions in which they were involved.

Learning in Church and Learning in School

At one point while interviewing Tia, I asked her to compare learning in church with learning in school. She exclaimed, "Church is more interactive!" She went on to explain:

> The teacher kind of puts a question or statement out there and you respond. And then it's like, you're reading the Bible out loud and it's like, "What do you think about this? What do you think about that? What do you understand from this? What do you understand according to what is being put out there from the word of God?"

Tia described an experience of sense-making that requires her to actively engage in thinking and understanding the material, rather than simply accepting another's interpretation. The teacher's initial comments, putting "a question or statement out there," did not seem to interfere with Tia's ability to form opinions or build understanding. To the contrary, she took this initial position as a call for her to become a more active learner; she was supposed to ask, "What do you think about this?" and "What do you understand from this?"

Tia seemed to embrace the feeling of being able to interpret something and the notion that different interpretations and different people's interpretations (even when there are power differentials between the people) are all possible and can contain truth. Tia also seemed to be commenting about the nature of mathematics in these remarks, drawing an inherent distinction between the natures of the two bodies of material. Whereas earlier she had described mathematics as "step-by-step," she stated that "sometimes the Bible isn't specific"; there is room for interpretation. The pastor may put forth a theory or a specific way of reading a passage, but it is just that. The fact that Tia saw the Bible as not "specific" allowed for different reactions and interpretations.

Tia used the word *interaction* at other times in the interview, in describing an effective history teacher and when asked how a mathematics class should run. Tia said that a mathematics teacher should explain the concept and then ask small groups to wrestle with it.

That is still interactive math, because I'm interacting with my group, and we're trying to figure out the problem. And then we present it as a whole group, and then questions may arise or questions may not arise and then, you know, I get to see if we got the right answer, or what did we do wrong, things like that, so that's still ... interactive learning.

Not only did Tia state that she learned best through interaction, but her notion of interaction did not preclude a teacher explicitly showing the steps first. In the model that Tia put forth, a teacher's demonstration complements, and perhaps primes the student's interaction with the mathematics, rather than interfering with it. Greg also described effective learning at church:

After the service is done, I always go home and just think about what he [the pastor] talked about. ... I just think about it, and just try to get an understanding of it ... of "How can I relate this to my life?" or, "How does this concern me?" ... When the pastor gives you what they perceive their message is, when he gives it to you, I think that you're supposed to, um, take it into your own situation, and relate that to your own life.

The sense of pliability and utility in Greg's description of the church's teachings again stands in stark contrast to the notions of mathematics as a fixed entity. Similar to Tia, Greg described a pedagogical middle ground. He said that "the word" is "given" to him, and that only after that does he take it home for interpretation. The pastor does not simply pose a hard problem and ask the congregation to puzzle it out. Rather, while the pastor might begin by posing a problem, he goes on to say what he thinks about it—where his learning and thinking has led him—and then expects people to take it into their own lives. Pastors model for their congregations how one might wrestle with a problem, and in so doing give what they "perceive" the message to be. This situation allows both for pastors to give their full interpretation first and for each member of the congregation to take it into his or her "own situation." There is no conflict presented between explicit teaching and individuals constructing their own learning.

In a comment that recalls his preference for one-on-one time with the teacher, Greg said that church is

an environment where 99% of the time you're safe, you feel safe. ... You just relax ... just leave all the negative things out and just put your full attention on what he's saying and then take that back into your ... just use it into life.

These depictions of church—safe places in which one is expected to make sense of the material—sound very different from the typical urban, public high school.

What Should We Make of These Students' Comments?

The students' comments about church strike me as contributing to the picture of a hybrid pedagogical style that combines teachers telling and students wrestling with the material themselves; they describe situations in which they are encouraged to talk and listen to each other and to voice their questions after first being given ideas with which to interact. Tia shows that she understands and can articulate certain elements and benefits of a learning community even while she may not view mathematics class in that way. The students seem to be questioning the fit of the goals and approaches I have laid out for the mathematics classroom and suggesting other ways in which those might be realized. We would expect different contexts to lend themselves to different teaching styles, but the students' comments suggest the potential for overlap and the borrowing of practices. It seems clear that we need to continue to learn about other institutions our students are involved in and what they can teach us about effective teaching.

CONCLUSIONS

This research reported in this chapter attempted to investigate some beliefs and attitudes toward the nature of mathematics and its teaching and learning. The outcome of such an investigation is not a formal model but a type of cultural commentary—to raise questions, undermine stereotypes, elicit possibilities. The temptation to draw simplified conclusions, for example, to summarize the work in findings about "how to teach," does a disservice to what it is that teachers actually do, and how we conceive of that task. The students in this study raise and complicate a number of important questions. They push us to rethink the nature of interpretation, the supposed dichotomy between direct instruction and constructivism, the roles and relationships of algorithms and reasoning in mathematics, and the utility of mathematics in their lives.

When I embarked on this research, my stereotype of church culture leaned toward "explicit," while I viewed my own teaching as "constructivist." The students described a church culture that is "more interactive," in which they were supposed to interpret the teachings. They view interactive mathematics not as pure discovery but as a hybrid—a negotiated middle ground—where students respond to as well as generate ideas. There were times when they wanted to be told, and others when they wanted to puzzle things through for themselves. This analysis leads to a number of critical questions for the research community, such as: How is interpretation in mathematics similar to and different from other academic fields and cultural contexts? To what extent is there a place for personal interpretation in mathematics? In what ways are students able to take mathematics back into their lives, as Greg does with what he learns in church?

I would argue that teachers need to be active and ongoing participants in investigating these research questions, both at the professional level— with other teachers and researchers—and with their students. In addition, we need to push ourselves to consider in-depth questions about race and culture that may be uncomfortable, but that lie at the critical boundary between the current state of our schools and our ideals. The point of this research was to develop increasingly rich conceptions of the landscape: to build our understandings of the role of culture in the various interactions in the classroom, and to use these understandings to inform the instructional decisions we make and recommend. Shulman (1986) describes pedagogical content knowledge as "the ways of representing and formulating the subject that make it comprehensible to others" (p. 9). Perhaps these students are pushing us to develop our *cultural content knowledge*—how we represent and formulate mathematics to make it comprehensible to others, *particularly across cultures*. To do this work well—to understand more fully the teaching of all children, and specifically children of color, to the highest levels—will require continued examination of these questions.

ACKNOWLEDGMENTS

I would like to thank The Spencer Foundation for its support of the teacher-research group through which I was able to start this work. I am grateful to Ricardo Nemirovsky and the people at TERC for their many years of generous support and feedback.

NOTE

1. The teacher-research group was created with support from the Spencer Foundation's Practitioner Research Communication and Mentoring Grants Program. The grant allowed us to hire outside expert facilitators to run monthly meetings in which teachers were supported to develop and investigate questions about their own practice. At the time, I was lucky enough

also to be collaborating with a research group at TERC looking in part at my classroom primarily through extensive videotape analysis.

REFERENCES

Ballenger, C. (1998). *Teaching other people's children: Literacy and learning in a bilingual classroom*. New York: Teachers College Press.

Boaler, J. (2002). *Stanford University mathematics teaching and learning study: Initial report—A comparison of IMP 1 and Algebra 1 at Greendale School*. Retrieved from www.stanford.edu/~joboaler/Initial_report_ Greendale.doc

Delpit, L. (1986). Skills and other dilemmas of a progressive black educator. *Harvard Educational Review, 56*(4), 379-385.

Delpit, L. (1988). The silenced dialogue: Power and pedagogy in educating other peoples' children. *Harvard Educational Review, 58*(3), 280-298.

Delpit, L. (1992). "Acquisition of literate discourse: Bowing before the master?" *Theory Into Practice, 31*(4), 296-302.

Delpit, L. (1995). Teachers, culture and power: An interview with Lisa Delpit. In D. P. Levine, R. Lowe, R. Peterson, & R. Tenorio (Eds.), *Rethinking schools: An agenda for change* (pp. 136-47). New York: New Press.

Ernest, P. (1996). Varieties of constructivism: A framework for comparison. In by L. P. Steffe, P. Nesher, P. Cobb, G. Goldin, & B. Greer (Eds.), *Theories of mathematical learning* (pp. 335-350). Mahwah, NJ: Lawrence Erlbaum.

Freire, P., & Macedo, D. (1995). A dialogue: culture, language and race. *Harvard Educational Review, 65*(3), 377-402.

Gallas, K. (1994). *The languages of learning: How children talk, write, dance, draw, and sing their understanding of the world*. New York: Teachers College Press.

Hiebert, J. (1999). Relationships between research and the NCTM Standards. *Journal for Research in Mathematics Education, 30*(1), 13.

Moses, R., Kamii, M., Swap, S., & Howard, J. (1989, November). The Algebra Project: Organizing in the spirit of Ella. *Harvard Educational Review, 423-443.

Moses, R. P., & Cobb, C. (2001). *Radical equations: Civil rights from Mississippi to the Algebra Project*. Boston: Beacon Press.

National Council of Teachers of Mathematics. (1989). *Curriculum and evaluation standards for school mathematics*. Reston, VA: Author.

Ochs, E., Taylor, C., Rudolph, E., & Smith, R. (1992). Storytelling as a theory-building activity. *Discourse Practices, 15*(1), 37-72.

Shulman, L. (1986). Those who understand: Knowledge growth in teaching. *Educational Researcher, 15*(2), 4-14.

Walker, E. V. S. (1992). Falling asleep and failure among African-American students: Rethinking assumptions about process teaching. *Theory Into Practice, 31*(4), 321-327.

CHAPTER 11

TEACHING MATHEMATICS WITH PROBLEMS

What One Teacher Learned Through Research

Nicole Garcia
Washtenaw Technical Middle College

Patricio G. Herbst[1]
University of Michigan

One key notion that mathematics teachers get from the NCTM *Principles and Standards for School Mathematics* (2000) is the value of using interesting and worthwhile problems to teach mathematical concepts to their students (see also Lampert, 2001). Unfortunately, finding worthwhile problems that one can use in teaching may not be as easy as it sounds. A worthwhile problem is not simply a problem that the students may see in their everyday life— it is a problem that gives them a fair chance to think about and explore mathematics. If a problem is to be usable for teaching something new, it also must push students to think beyond what they would do spontaneously.

Teachers Engaged in Research
Inquiry Into Mathematics Classrooms, Grades 9–12, pages 197–214
Copyright © 2006 by Information Age Publishing
197

Some of these problems may lead students to work and think in novel ways. They may even require students to examine ideas that they later discover are completely wrong, but whose consideration illuminates a path towards a correct solution.

As a mathematician, I realize the importance of examining and working through an incorrect solution to a problem in order to build the correct solution. However, as a teacher, it is a challenge for me to use problems in my teaching that might lead students to work through an incorrect solution. It is a challenge because I realize that from the students' perspective I represent a trusted mathematical authority—no matter how much I might like to make students responsible for becoming agents of their own learning. In this role, I feel responsible for providing students with clear, correct information; for guiding them to correct ideas and not leading them astray. Asking questions that introduce incorrect ideas as a step toward correct ideas can force students to think and evaluate ideas, and it also gives them a sense of the nature of mathematical work. Still, I wondered how far a teacher should go with these questions and how a teacher could make them work for the students as well as for herself. My participation in a classroom research project both raised these questions and gave me the opportunity to struggle with finding answers to them.

BECOMING INVOLVED IN A RESEARCH PROJECT

I was first involved in teacher research when I student taught during my final semester as an undergraduate student in mathematics education. Research is not a part of the undergraduate curriculum at the university I attended, but during my pre-student teaching semester in the fall, my mathematics education professor, Patricio Herbst (Pat), began to visit the classroom of Megan, the cooperating teacher I would work with as a student teacher in the following semester. Megan had taught algebra and geometry for approximately 12 years at the time that I first met her. She was used to a teacher-directed teaching style but had agreed to be a part of a research project directed by Pat that would involve students in more independent work. She was interested in participating in a research project out of her wish to learn new things and because she valued showing her students that other adults cared about what they thought. Whereas I respected the choices Megan had made regarding how to conduct her classes, I was looking forward to learning how to create a classroom environment in which students had lively discussions about mathematics and felt comfortable expressing their views and concerns. In my view of an ideal class, I wanted students to ask lots of questions and start conversations.

The project, as I understood it then, consisted of developing lessons on area and evaluating how these lessons affected students' ideas of the process involved in proving and what it means to do a proof. Based on my own experience as a student and my observations in geometry classes since then, I concluded that most high school students think of proof as a two-column checklist of items they have memorized that show why some statement is true. The statements that they proved were traditionally not their own conjectures, but known-to-be-true things that had been stated for them to prove. Thus, even though they knew the statements were true, they had to prove them again for no particular purpose except, as one student would say later to me in interviews, "to learn how to be good thinkers." The lessons developed for this project put students to work on problems where proving was actually helpful in finding and justifying a solution.

I found the concept of area to be an interesting topic for this study, because area seemed to be generally taught by giving students formulas and having them practice applying those formulas. I always wondered why there weren't really any proofs involved in area, only straightforward applications-based problems. I have experienced and observed area being treated more like algebra with numerical problems than like geometry with proof problems. Because of this, I thought area would provide an interesting examination of students' ideas of proof and the uses of proof.

My Role in the Research Project

My initial role in the research project was to help create lesson plans independent of the textbook series used in the classroom, but covering the same material. Pat had a list of problems that he thought could be used and also conveyed his expectation that solutions to those problems be justified. He worked with Megan and me, weeding and sequencing the list of problems. Then Megan and I planned with relative independence how we would conduct each lesson and how we would connect the solving of each problem to the material that needed to be taught. For Megan it was important that the lessons covered topics on area that would be assessed in standardized tests, such as the area formulas which were part of the prescribed curriculum. Other than that she welcomed the opportunity to give her students a change of pace by having them work on challenging problems.

As illustrated in the next section, each lesson was developed around a problem for students to work on in groups of four during their 55-minute geometry period. The problems required the students to justify their answers, essentially giving a proof, but not the formal, two-column type that students are usually expected to give. They involved critical thinking, transfer of prior knowledge, and a certain degree of assumption—something oftentimes not allowed in mathematics, but required in life just to make problems wieldy. Examples of such problems are the central focus of the remainder of this chapter.

I helped design the lessons so that there was no explicitly "correct" answer and the strength of one's answer lay in the justification. Students were asked to describe their reasoning for all classwork and homework, both in the small group setting and in the whole class setting. They were consistently questioned about the justification of their answer, whether the answer was or was not appropriate. Part of the goal of the study as I saw it was to examine, and possibly change, students' views on what it meant to do mathematics and what constituted a proof. After conversations with Pat, I came to understand that he was interested in seeing to what extent problems that demanded students say things about a figure that were not perceptually evident could elicit from students the kind of mathematical reasoning one usually uses in proving (see Herbst, 2005).

The second goal of the study, I learned as the project unfolded, was to examine from the perspective of the teacher the feasibility of making proof appear as a tool for finding out a solution to a problem in a classroom. Among other things, Pat was interested in the decision-making process of a teacher as she balances allegiances to what is mathematically interesting and sound and to what her students need, given where they are at. As a novice teacher, it took me a while to realize that this was part of the work of teaching—making decisions that often were not clear cut but that involved managing dilemmas (see Chazan & Ball, 1999; Herbst, 2003, in press; Lampert, 1985).

One of the reasons this realization took some time to develop is because I was still in the role of a student transitioning to the role of teacher. In my initial evaluation of the project and the work we were giving students, I had taken on the view of the work as a student and tried to imagine the students' opinions of the lessons. Instead of immediately seeing the value and benefits of the types of problems we were giving as a tool for learning, I visualized the resistance that I, as a student, would have given in response to the problem. I examined the directions given for homework and those stated in the lesson plans and imagined all of the objections that could come from students—"What am I supposed to do?," "Why aren't there any numbers?," or "You haven't taught us this yet so we can't solve the problem." It was from this perspective that I began the teaching part of the project.

I was responsible for teaching the lessons to one of our three geometry classes—Megan taught the other two. Each of the twelve lessons of this special unit on area was videotaped in all three of our geometry classes. In addition to videotaping the lessons, data were collected through individual interviews of selected students throughout the semester; collections of all worksheets, homework, and assessments; and photocopies of the notebooks of interviewed students. Videos and student interviews were then discussed by a group that included Pat, graduate students, Megan, and me. While viewing videotape, the group examined, among other things, the rationale for the decisions the teachers made as they asked questions, supported students, and accepted/rejected student answers, and how the objectives they had in doing that matched student expectations. Student work was used to enhance this discussion.

Teaching With Problems

Prior to implementing the research problems, our classes were conducted in a manner typical of U.S. classrooms (Stigler, Gonzales, Kawanaka, Knoll, & Serrano, 1999). The beginning of the class period was dedicated to answering questions from the homework. This was followed by notes, classwork, and the assigning of homework. During note time, students were involved in teacher-led whole-group discussion to help develop the concepts being discussed. The students were prompted by the teacher to answer questions she asked them and were allowed to ask questions of the teacher. These discussions dealt with traditional scaffolding problems that built directly upon each other to help complete a final problem or idea. Each question was carefully chosen before class to examine common misconceptions or to show students how the mathematics would be used on the homework they would be doing. Questions that built upon each other did so by presenting the need for an additional skill, which was taught after the presentation of the problem. I understood Megan's reasons for structuring her classroom in this manner. She had 55 minutes each day to cover the material from the district curriculum. Her planned example problems allowed her to deal with common misconceptions, thus reducing the risk that one of the students would solve a problem incorrectly. Her scaffolding problems enabled her to get through the lesson by the end of the class because they led students to the correct solutions of problems. As I thought of my ideal classroom, however, I wanted more participation from students, more discussion, and generally a more student-centered classroom.

These traditional lessons took place for approximately six months until Megan and I began to teach the special lessons on area. The replacement unit on area lasted for about three weeks and focused on solving "big problems," such as comparing the area of a quadrilateral with the area of the quadrilateral formed by its midpoints. One big problem in the unit was an area problem borrowed from a Japanese lesson from Stigler and Hiebert's (1999) video study of mathematics teaching in various countries. The problem asked students to figure out how to straighten a crooked fence between two neighbors' properties without changing the area of each neighbor's property. We used this problem to bring out the idea that triangles that share a side and have a vertex along a line parallel to that side have equal area. Furthermore, the problem was meant to impress upon students that one can say some things about the relationship between the areas of two figures even when one does not know the specific areas of those figures.

In the above problem, students were expected to interpret the meaning of the problem, possibly create some examples, and work through several possible solutions. The end result would be a triangle of equal area drawn using an auxiliary parallel line and a justification that this new fence does not change the area of the neighbors' respective properties, even though we don't know that area. The formulation of the problem differs greatly from traditional proof problems in that students must create their own "given" and "prove" and they are given no hints or guidelines as to how to complete the proof. They are also not explicitly asked to give a proof. However, this situation represents more closely the way that proof would present itself in the work of a mathematician or anyone who uses mathematics. Still, these problems made me feel uncomfortable as they did not make expectations of the type of solutions they are expected to produce clear to students.

Another big problem of the area unit was what we have been calling "the triangle problem" (see Herbst, 2003). It was this problem more than any other one that brought about my misgivings about asking students to consider an idea that would eventually be shown wrong. Students were presented with the following question: *Say you have a triangle ABC, how can you find a point O inside ABC so that the three new triangles AOB, BOC, and AOC have the same area?*. The work on the problem was meant to take two days and to conclude with the students giving instructions for how to find such a point and proving that the resulting triangles had equal area. Since we expected they might not easily move from the problem to its justified solution, we had prepared some intermediate problems that we could use as a scaffold. For example, Pat thought that students might need some help realizing that what the problem was really asking was to specify a construction; furthermore, finding what construction to specify was not obvious either. So he proposed giving students a homework problem that dealt with a potential solution to the problem that we called Last Year's Conjecture.

LAST YEAR'S CONJECTURE AND
A PROBLEM OF TEACHING

Last Year's Conjecture asked students to consider a suggestion purportedly made by former students about how to solve the triangle problem. These-

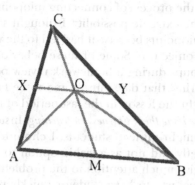

Figure 11.1.   Last Year's Conjecture—
AOC, BOC, and *AOB* are equal in area.

former students had supposedly suggested that the solution to the triangle problem was to connect the midpoints of two sides of the triangle and then find the midpoint of this segment; this point would solve the problem. This task certainly had some pluses; for example, it conveyed a sense of what a solution could look like and it gave students a way to think about "the middle" of a triangle in a more operational way. More importantly, the conjecture was false, but the ways in which students might show that it was false could help them figure out what the correct conjecture would have to be like. Yet I did not like this homework task one bit.

What was the issue? In the initial planning of the lessons and our later inspections of the video clips, I had many misgivings about Last Year's Conjecture. My concern was about the fairness of asking the students this particular question. I felt that the question misled students. The school norm that is instilled in children beginning in early elementary mathematics is that students are given a problem as a scaffold to help them solve another harder problem. While the conjecture problem was technically a scaffold to help students solve the triangle problem, it differed from traditional scaffolding problems. Traditionally, the solution and methods used in the scaffolding problem are used in subsequent problems. The conjecture problem, however, did not do this; rather, it showed students what *not* to use in subsequent problems. Giving this problem as a scaffold for their work contradicted my understanding of my role as their teacher.

It appeared to me that students inculcated in school norms could be misled by this problem, assuming that because the teacher was giving this conjecture to them for consideration, the conjecture was correct and it was a hint to help them solve the triangle problem. I imagined that students would focus too much on the problem, assuming that even if it wasn't the exact point needed, the process of connecting midpoints must be involved in the solution. In the extreme possibility, I thought some students might even dismiss using midpoints because it had led to the wrong answer in the case of Last Year's Conjecture. Some of these behaviors were, in fact, witnessed in several groups during a homework review period in the lessons taught by Megan earlier that day, causing me to be faced with a difficult decision when I taught the lesson in the last period of the day.

How I dealt with the issue then: Fairness in teaching. Instead of using a problem that I worried might mislead students, I chose to discuss Last Year's Conjecture very briefly and not as a viable option to solving the triangle problem. I did not give much attention to the problem as a suggested possible solution; my goal was to have students quickly recognize that it did not work. Every group had seen the problem in class, but I only discussed it with one group. The other groups did not spend much time considering the problem as they were fixated on the solutions they had created in the previous session. Here is what I said to the group composed of Alan,[2] Anna, Christian, and Cory:

> Let's turn it over and let's look at the problem in the back. This is just an example that another student came up with that they thought could be an answer and I just want you guys to draw the picture that they are asking you to draw and then tell me whether you think that is possible or is not possible, or if it's gonna work with just equilateral triangles or... Okay?

Part of the reason I approached it in this manner and with only one group of students was that the students had spent a day working on the triangle problem and already had an idea of where the equal-area point should be—in fact they had what, in my mind, was the right idea. Unfortunately, though the students were heading in the right direction, they were choosing their solutions only from points they had learned earlier in the semester without ideas about why this point should be the solution. I felt I had to do something to enable them to figure out why the centroid was the desired point. However, I was not sure the correct move was to give them Last Year's Conjecture, because I felt that misleading them at this point in their work would be unfair to them. I thought doing that would give them the impression that they were on the wrong track, which they weren't, and might lead them in an unproductive direction.

My misgivings about the fairness of Last Year's Conjecture were in line with the views of the students. During interviews of several students over the course of the research, I found that students saw a teacher's problem choices to be based on fairness. They thought a problem was fair if it gave enough information to be done, required skills and concepts learned recently in the current unit of study, and was stated clearly. Hearing them speak about problems was like hearing an echo of my own comments about Last Year's Conjecture.

Even the statement of the problem could be construed as unfair. The problem was said to be a suggested solution from another group of students. From the point of view of a student in the geometry class, a teacher shouldn't mention students' incorrect answers in front of the class, as this can be seen as demeaning and embarrassing. Because of this convention, students might assume that part of Last Year's Conjecture, if not all of it, must be correct. On the other hand, I could also see how stating the conjecture as made by some anonymous students might be better than presenting it as a teacher's idea; students would then feel free to criticize and examine the problem without worrying about offending the teacher.

In spite of the fact that I gave almost no time for them to think about Last Year's Conjecture, in my view at the moment, the overall lesson still had wonderful results. My goal was not to sabotage the students' learning, but to help them get the most out of their work. The groups eventually solved the triangle problem and shared their solutions with the class. One group from this section gave a wonderful algebraic proof with an additional piece showing that all six smaller triangles formed by drawing in the medians had equal area, in addition to the three larger triangles that they were supposed to find. These students created this proof without the scaffolding that was supposed to be provided by Last Year's Conjecture.

My own perspective on the issue of fairness in problems started to change during discussions with the research group. Other people began to defend the problem and highlight the benefits they saw in giving it to students. One such defense for giving students this type of problem was "you could think, well, the right solution, the good thing about the right solution is it allows you to think why this one is wrong. And giving people exposure to a wrong solution makes them appreciate the right solution." They conjectured as to what they thought could happen when using the problem. An idea was that "you progress ... to a more reasoned right answer and one way I thought that could happen was by giving them a wrong answer that they could critique." They dissected student comments and work to see what levels of understanding were represented in those groups that had considered Last Year's Conjecture as compared to those that had not considered the problem. I started to consider whether there were benefits or reasons for a teacher to give a problem like Last Year's Conjecture—a problem that would use an incorrect idea as a platform from which to take the students' investigation to the next level—in this case that of discovering what had to be true about the correct solution being sought. Through our discussion, I began to realize that these problems did have value in the classroom. Although I was satisfied with the understanding demonstrated by those students who did not consider Last Year's Conjecture, I became convinced that the students might have had an even deeper understanding if they had the chance to consider this false conjecture. I knew that while students were learning the concepts without Last Year's Conjecture, I was doing them a disservice by not providing them the additional depth of understanding provided by examining why the conjecture may or may not work.

As a result of our research group discussions and comparisons between the quality of the mathematical work that students had done before and during the research lessons I began to understand that problems like Last Year's Conjecture might afford students a chance for independent thinking. Students are not given a set algorithm to solve the problem, but rather must create their own framework for interpreting and solving the problem. They are also allowed to share their ideas with the mathematics community, since there is no one right way to get to the answer. Usual scaffolding problems provide a false sense of independent thinking; they really tell students where they needed to go next rather than allowing students to create their own path.

Problems like Last Year's Conjecture likewise serve to assist the teacher in her work. No longer is the teacher the sole source of mathematical knowledge—students are now contributors to the classroom mathematical discourse. The classroom is enlivened with discussion when students have the chance to argue for and against a proposed conjecture; this can be more valuable than whether or not the conjecture itself is correct. Deep understanding of the concepts is formed through the debate over the adequacy of someone else's ideas. This allows teachers to cover the topics students need deeply and in a timely manner.

I realized that Last Year's Conjecture was unfair only from the perspective of the traditional student. From the perspective of a teacher focused on expanding students' understanding of proof, it was a valid question that allowed many learning opportunities. I have come to see that thinking like a teacher requires not only allegiance to what students want but also allegiance to the mathematics they are supposed to learn and the best ways a teacher knows how to help them learn it.

The next step for me was to find out how I could use problems like Last Year's Conjecture in the classroom while maintaining a strong student–teacher relationship based on a sense of trust and fairness. I now see my role as a mediator between the students and their work. I need to take on the perspective of both the students and the mathematics expert who sees the benefit of working through tough problems without hints. I also need to provide the students with a sense of why the problem is important for them to solve, as a mediator resolving arguments would.

MY OWN CLASSROOM:
RUNNING INTO THE PROBLEM AGAIN

I took a job as a full-time teacher the year after my student teaching. The context of my teaching was very different than that of the project classroom and made issues of fairness even more important. The original study took place in a large, comprehensive high school within a district that spent more than $9,000 per student and whose graduation rate was more than 90% in 2002. The district was fairly racially diverse with a student body that represented many minority groups, and the classes I worked with each had about 24 students who were all in the regular college-preparatory track (Standard & Poor's, 2003).

The much smaller charter school in which I teach now has a similar racial makeup, but spends half the money per student and takes students from various districts, many from blue collar families. My school's maximum class size is 20 students, and the students come from a variety of educational backgrounds. Some students have failed out of their old schools, some have performed at a mediocre level, and others excelled. All of them chose to attend my school, mainly because they disliked their previous high school experience. These students of varying motivation and abilities are in the classroom together. Due to their various prior school experiences, it is important to make the students feel comfortable and to establish trust with them. I would argue that this is important in every class, but especially in high school and middle school mathematics when many students start to feel very self-conscious, others just want to be finished with school, and mathematics may no longer be a required course. High school is also the time when many students take algebra. In my experience, I have found that many parents believe that some people just can't do algebra and as long as they pass the class, their student is done with math. Because of this, it is important for me to think from both the perspective of the student and of the teacher.

MY SOLUTION: MAINTAINING ALLEGIANCE
TO BOTH MATHEMATICS AND STUDENTS

During research group discussions about the fairness of this type of scaffolding problem, I began to think about how both parties could be satisfied—how teachers could give their students quality problems that allowed students to be in charge of their own thinking and learning and still be perceived as being fair to students. I arrived at the conclusion that since students' ideas of fairness were the product of classroom norms, the precedent would have to be set early in the year for the types of problems we would be doing and the expectations for student work. Part of this process was to introduce problems that, according to students' prior experiences, might be viewed as unfair and, after their use, to divulge and discuss my reasons for using them with the students.

In my own classroom, I have attempted to implement lessons similar to those used in the research project. I began setting the classroom norms by making it clear to students on the first day of class that no ideas would be presented without some sort of justification. This was supported throughout the year by consistently asking "why" questions when students made conjectures or offered solutions. Also, to keep students from feeling that my "why" questions were a form of criticism or a hint that their answer was wrong, I consistently asked every student to justify his or her answer and I qualified my questions with statements such as, "I am not saying that you are right or wrong, but I would like to know..." instead of simply asking them the question "Why?"

The types of problems I used to scaffold my own students' thinking reflected the properties of those used in the research project. They were not traditional scaffolding problems that led directly to a solution. Instead, these problems required students to make conjectures and back up their claims.

An example of such a scaffolding problem used in my classroom consisted of an activity that I intended to use to help students build and prove a formula for determining the sum of the interior angles of any shape. I began the discussion by posing the following problem to my students:

> Okay guys. So we know that all triangles have angles that add up to 180 degrees. But what about quadrilaterals? I have had students tell me that the sum of the angles in a quadrilateral can vary, depending on what the quadrilateral looks like. I want you to think about whether or not you agree with that statement. A good way to start might be to draw a few and estimate what the angles are.

Asking them to draw a few gave them a starting place and gave some validity to considering the question by requiring everyone to contribute something to the problem. After a few minutes working in their groups, one of the students, Omar, raised his hand to object to my claim. "We know a quadrilateral is 360 because it is made of two triangles," he claimed. I countered his statement with the following drawing:

I made the counter-claim that his statement was not true because "I just drew a quadrilateral with four triangles, which gives us 720 degrees." I asked students to share with each other what they had found in their

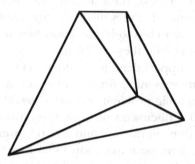

Figure 11.2.   Another way of breaking a quadrilateral into triangles.

groups. All had found quadrilaterals with about 360 degrees of interior angles. I then posed the question:

> If Omar is right and the angles of a quadrilateral are 360, how can I make a rule about how to figure that out using triangles? I think we need to make a rule because I have drawn a quadrilateral with lots more than 360 degrees worth of angles.

Following some discussion among the groups, students came to the realization that only triangles whose angles were part of the vertices of the quadrilateral and did not overlap each other could be used.

Before the discussions and observations I took part in during the research project, I would have seen this problem as unfair. The students were obviously on the right track in determining that quadrilaterals' interior angles added to 360 degrees. I would have seen it as my duty to reinforce these ideas by giving the explanation myself or simply asking Omar how he came to this conclusion. I also would have seen giving the students a false statement as problematic, since many students listen only part of the time and I am responsible for ensuring that they understand the material. Giving a false statement would be like feeding them the wrong information. However, after having spent much time mulling over fairness and unfairness to students in working with problems, I saw that a statement such as the one about four triangles making up the angles of the quadrilateral is a valuable tool to make sure that students consider all possible options and give a strong justification for their solution.

When I promote evaluating conjectures these days, I notice myself helping the students to make more connections, because I can build on the ideas that they raise. Those students who tell me at the beginning of the semester that they are "not good at math," and generally would want to lean back and let others take the lead, actually engage in the discussion and take pride in the fact that they made a contribution. Many of these students approach me after class to let me know how much more they learn and understand when we do these problems.

A critical piece of support that I have added to my own lessons is sharing with the students my reasoning for assigning certain kinds of problems. Upon completion of a big problem that used several scaffolding subproblems, I usually hold a discussion with my class. We talk about all of the problems that were done in the unit and how they were connected to each other. Students give their ideas about why I may have chosen the particular problems and then I share my reasoning. I feel that this allows students to understand that I consistently consider what work I ask them to do and that I always have a purpose in mind. Surprisingly, this helps my students in another way. Instead of just doing whatever work I give to them, my students try to make connections between the problem at hand and different problems they have done before; as a result this has improved their transfer. For example, after considering Omar's approach to the quadrilateral problem mentioned earlier, students built the general formula for the sum of the interior angles of any polygon. This contrasts markedly with previous years when many students memorized the sum of the interior angles of different polygons instead of seeing the pattern.

PARTICIPATING IN RESEARCH AND IMPROVING MY TEACHING

The classroom experience in the research project was the backbone of my learning to teach with problems, but other aspects of the research also helped me in my teaching. The research provided me with insight into the students' expectations of how problems would be chosen. It gave me confidence to experiment with negotiating those student norms in my classroom. In the beginning, it was quite challenging to get students to even attempt problems they thought to be unfair. It was a slow process of countering the students' expectations of the problems they would be given with my reasons for asking them to do these types of problems. Once the students saw that mathematics did not usually present itself in real-life situations in the same way that it was presented in traditional textbooks and classrooms, many of them were willing to make the effort to at least try the problems. This would not have been as easy for me to deal with had I not had access to the students' perspectives as I did with the interview transcripts. These gave me insight into student opinions and expectations. Throughout the years of data collected, several student comments stuck with me.

In particular, student comments on the purpose of proof in geometry have been helpful. For example, when asked about whether proofs are ever needed to know whether something is true, a student said "in some cases you can see that something is true without doing the proof, but there have been times where it didn't seem true until you actually solve it." Another student noted that doing proof "you get like a more logical thinking process." Others commented on the difference between the problems they were asked to solve in the unit and those they are usually assigned, saying things like "the proofs we [usually] do are more straightforward than this and they're not as open-ended" and "we tend to do proofs at the end after we figure out some fundamentals about some theorems that we've learned from the chapter." Many found proofs to be pointless in cases where they already knew that what they were proving was true. In such a case, one student stated, "if it's an assignment I'll actually solve the proof, but if it's a real-life case then it doesn't really seem, there isn't a point."

As teachers, we would like for our students to understand the importance of proving conjectures rather than taking statements at face value. Many times it seems to us as though we repeatedly explain to students the significance of proof and justification, and we support this by providing examples and problems of real-life applications. However, without student feedback, it is hard to determine whether they are receiving the message we intend. The student interviews helped me to see that while students frequently repeat the words we use to justify learning and doing proof, they may not actually synthesize the information. Many students interviewed stated that proof was an important life skill, but when asked what subjects it could be used in, some failed even to recognize that it was useful in algebra, let alone real life. One student, when asked why she thought she had done proofs in geometry class, but not algebra stated "Um, well in algebra you're just doing it with numbers I think and it's easier to prove without a proof ... they're more like ideas I think more than like it just happens." This strengthened my belief that teaching through problems was a way to help students understand the power of proof. It seemed to me that the traditional methods of teaching were not helping students see this, but using open-ended problems that required justifications might.

The creation and implementation of these independent lessons provided another excellent experience that influenced my subsequent teaching. When I first began my student teaching, I thought of the textbook as my main resource for lessons. I made the assumption that the authors of all textbooks ensured that appropriate questions were included in the text and the teacher's edition so that students developed a deep understanding of the mathematics, as well as critical thinking skills. Creating and teaching the lessons developed in this study allowed me to see the difference in what could be provided to the students if I introduced more open-ended problems rather than simply using problems from the book chosen by the school district. While this may not be a problem in all schools—depending on the goals and format of the textbook—it can be an issue in many districts that require teachers to work within the confines of an adopted textbook that doesn't reflect the teachers' goals for their students. Developing additional materials allows teachers to satisfy the district's requirements as well as their own goals for their students.

Before the research project, I had visions of myself teaching my class much as the classes I had taken as a student. I assumed I would give notes and ask students conceptual questions, but most of the work I would give them would consist of straightforward computational or standard proof problems. Instead, I now teach using interesting problems that allow students to explore, share solutions, and discuss concepts and possibilities. Being involved in a classroom research project allowed me to explore different options for teaching mathematics—options I had not considered before.

Being involved in a research project has lent credibility to changes in my classroom that might have otherwise been difficult to implement because they differed from the school norm in the eyes of parents, colleagues, and administrators. It also took some of the pressure and responsibility off of me as the classroom teacher as I was not solely responsible for the implementation of the methods or the creation of the alternative materials. I had the support of an expert in mathematics education to guide me in both areas. This allowed me to try new and adventurous things that I wouldn't have felt comfortable doing alone.

The discussion sessions following each lesson and video viewing, as well as the occasional interviews with students, allowed me to engage in self-reflection and evaluation that I might have otherwise overlooked during the usually busy school year. While self-reflection is repeatedly recommended in teacher training programs and professional development courses and books, it can be difficult for teachers to find a starting point, to know what kinds of things make for productive reflection, and to assess the significance of their responses to the reflection. Involving oneself in a research project can guide the thought-process for reflection and can bring to light many details one may have thought of as insignificant.

These lessons and discussions provided a specific focus for reflection. They allowed me to zero in on one aspect of instruction—the evaluation and creation of problems for use in the classroom—and to learn how to reflect on this as both a teacher and a student. They also allowed me insight into other teachers' and experts' perspectives as to what one should be thinking about when choosing and using problems in the classroom. We spoke at length about issues of fairness of certain problems and how what might be perceived as unfair to the student may actually end up being good for them in the long run.

CONCLUSION

I feel that this research experience helped me become a better teacher—one who feels confident in teaching mathematics through problems. Similar experiences could be used to enrich teacher preparation at all levels and to allow teachers to explore new territory and gain new insights. Involvement in research might bring new life to a class that an experienced teacher has taught for years or help student teachers to gain perspective on the thought processes that go into planning a lesson. Research experiences provide a forum for discussion of issues such as methods, fairness, and depth of understanding that may be different from the department-wide discussions in which most teachers are involved. Most of all, allowing pre-service teachers to engage with experienced teachers in classroom research improves children's learning experiences by providing new teachers with the opportunity to expand their teaching repertoire and learn what it means to become a true reflective practitioner.

NOTE

1.   This chapter was written by Nicole Garcia with assistance from Patricio Herbst. The development of materials and collection of data referred to in this paper were supported by a grant to Patricio Herbst from the Office of the Vice President for Research of the University of Michigan. All opinions are the responsibility of the authors and do not necessarily reflect the views of the University.
2.   Alan, like all student names in the chapter, is a pseudonym.

REFERENCES

Chazan, D., & Ball, D. (1999). Beyond being told not to tell. *For the Learning of Mathematics, 19*(2), 2-10.

Herbst, P. (2003). Enabling students to make connections while proving: The work of a teacher creating a public memory. In D. Mewborn, P. Stajzn, D. White, H. Wiegel, R. Bryant, & K. Nooney (Eds.), *Proceedings of the 24th Annual Meeting of the North American Chapter of the International Group for the Psychology of Mathematics Education* (Vol. 4, pp. 1681-1692). Columbus, OH: ERIC Clearinghouse for Science, Mathematics, and Environmental Education.

Herbst, P. (2005). Knowing equal area while proving a claim about equal areas. *Recherches en Didactique des Mathématiques, 25*, 11-56 .

Herbst, P. (in press). Teaching geometry with problems: Negotiating instructional situations and mathematical tasks. *Journal for Research in Mathematics Education.*

Lampert, M. (1985). How do teachers manage to teach? Perspectives on problems in practice. *Harvard Educational Review, 55*, 178-194.

Lampert, M. (2001). *Teaching problems and the problems of teaching*. New Haven, CT: Yale University Press.

National Council of Teachers of Mathematics. (2000). *Principles and standards for school mathematics*. Reston, VA: Author.

Standard & Poor's School Evaluation Services. Retrieved September 27, 2003, from http://www.ses.standardandpoors.com

Stigler, J. W., Gonzales, P., Kawanaka, T., Knoll, S., & Serrano, A. (1999). The TIMSS videotape classroom study: Methods and findings from an exploratory research project on eighth-grade mathematics instruction in Germany, Japan, and the United States (NCES 1999-074). Washington, DC: National Center for Education Statistics.

Stigler, J. W., & Hiebert, J. (1999). *The teaching gap*. New York: The Free Press.

CHAPTER 12

REFRESHING MATHEMATICS INSTRUCTION THROUGH MOTION, TECHNOLOGY, AND A RESEARCH COLLABORATION

Apolinário Barros[1]
Boston International High School

Dorina Sackman
Stonewall Jackson Middle School

Imagine a classroom where students and teachers do not just study mathematics, but actually see it, hear it, even feel it as mathematical phenomena occur in their midst. Imagine students who can relate mathematical concepts to experiences in their own lives, because they have had a chance to experiment with physical embodiments of those concepts. Imagine a classroom where students and teachers work together to pose problems; articulate, develop and test solutions; and describe the inevitable surprising results they find. This chapter describes how the classroom described above became a reality for my students and me[2], and how my collaboration

Teachers Engaged in Research
Inquiry Into Mathematics Classrooms, Grades 9–12, pages 215–230

215

with the TERC Mathematics of Change Research Group[3] (MCRG) has influenced my practice and my students' ability to work together towards the common goal of mastering mathematical concepts through exploring physical motion with technology.

In my experience, there have been few opportunities in the public high school mathematics classroom for experimentation with motion as a way to understand formal mathematics. Rarely do high school students investigate physical phenomena as they study algebra, geometry, or trigonometry. Yet the investigation of real phenomena can be especially valuable in the study of these subjects, and indeed their applications to the sciences and engineering are widely touted in mathematical textbooks. Despite the strong connections of these subjects to the physical world, students are often expected to deal with ideas that are abstract and difficult to visualize without physical tools to help them experience these ideas more concretely. In particular, the National Council of Teachers of Mathematics (NCTM) states, "technology can be effective in attracting students who disengage from nontechnological approaches to mathematics" (NCTM, 2000, p. 14). For these reasons, I believe that including connections to students' experiences, physical motion, and technology in the high school mathematics curriculum is essential to reaching the maximum number of students and to helping them to understand the application of mathematics to the world.

This chapter describes how my collaboration with MCRG helped me to make these connections and, in the process, create more in-depth discussions and empower students to think for themselves. I have included an episode of classroom activity that exemplifies the senses of community and ownership developed by my students throughout this experience. I use this episode to illustrate what I learned about my teaching and the intellectual power of my students from my research collaboration with MCRG.

It is important to note that the results described in my opening sentences did not come quickly or easily. Being a mathematics teacher for the past 14 years in the public school system, I have participated in many professional development workshops designed to improve my teaching and the learning of my students. Although I feel professional development is essential to the growth of a teacher, oftentimes the professional development activities scheduled by the school system are one-day workshops with little feedback or follow-up for teachers attempting to incorporate the ideas of the workshop into their teaching. If these workshops were to be long-term, offer feedback, and involve classroom interaction, ongoing collaboration, and group discussions, they would be much more beneficial. In fact, my collaboration with MCRG bears out Stigler and Hiebert's claim that "Collaborating with colleagues regularly to observe, analyze and discuss teaching and students' thinking or to do 'lesson study' is a powerful,

yet neglected, form of professional development in American schools"
(NCTM, 2000, p. 19).

THE CONTEXT

MCRG has worked to bring the mathematics of change and motion into
mathematics classrooms from kindergarten to high school, with the goal of
making mathematics more accessible and relevant to students' lives and
experiences. Many of the activities created for high school classrooms
include a kind of technology developed by the project called LBM, or Line
Becomes Motion, which is a reverse of traditional MBL (Motion Becomes
Line) technology, such as motion detectors. While motion detectors allow
one to make a motion in front of a detector and see a graph of the position
or velocity versus time for that motion in real time, LBM technology allows
one to do the reverse as well. That is, with LBM you can make a graph and
cause a car or other device to move in accordance with the graph.

My work with MCRG began with a project called the Urban Calculus Ini-
tiative, in which a group of teachers from a public high school and a charter high school in Boston met monthly to reflect on their work as teachers
of mathematics, their classroom practices, their students' work, and their
own mathematics learning. The overall goal of this project was to work
together to increase the number of students in these high schools who
would have access to higher level mathematics classes, such as pre-calculus
and calculus. We planned lessons together and actually did mathematics
together to prepare for doing mathematics with our students. We created
and shared our classroom content objectives and activities. As time passed,
we wanted to explore in greater detail the teaching of a specific topic. This
led us to create physical activities through which our students could
develop a better understanding of the topic at hand. For example, one
such activity involved having students explore the idea of slope by physically moving a mini-car at different speeds and examining the resulting
speed-distance graphs.

Once we taught the lesson in our classrooms, we shared our experience
with the group using videotapes of the class sessions, audio-taped small
group discussions, transcribed dialogue, and samples of student work. Initially the purpose of this reporting-back was to get feedback and reflect on
this specific lesson, however, as I saw how the lesson affected my students'
abilities to better comprehend mathematics concepts, I became extremely
passionate about this integration of technology and "hands on" activities
in the classroom. Reflecting on data from the lesson allowed me to see an
amazing transformation in some students' mathematical and interper-
sonal skills.

As a result of my enthusiasm, MCRC researchers came into my classroom genuinely interested in observing my students to better understand how to help other teachers create a more dynamic learning environment. From a professional standpoint, being a teacher in high school can be a very lonely experience. I was reluctant at first to allow outsiders into my classroom, but soon welcomed these new faces to the classroom to refresh it, refresh me, and, most importantly, refresh the mathematical minds of my students. This was the start of a long-term collaboration where I joined staff from MCRC in analyzing videotape and other data from my classroom, trying to understand what was going on for the students as individuals and groups.

It was at this point that I began to see the imagined classroom described at the start of this chapter become a reality. There was more *talking* about mathematics than I had seen before and less simply writing the answer on the board. The students were asking more questions and these questions opened up the mathematical door much wider while *still* being aligned with the public school's curriculum. Basically, this new way of learning and doing mathematics got my students interested—really interested. It helped create a classroom environment in which students made the connections between different mathematical concepts—to me, one of the most important things for my students to learn to do. Mathematical concepts from previous lessons were being used as a tool to explain and justify new mathematical problems. This was exciting since one of the goals I have for my students is that they understand that mathematics can be learned through applications to everyday circumstances, instead of being something that they need to extract from me, the teacher who possesses the mathematics knowledge.

THE PHILOSOPHY

As a soccer coach, I often compare my classroom experiences with hands-on activities to my team learning and improving on the field. I asked myself, do you learn soccer just by watching it or reading a book about it? Even the best players need to get out there and feel the ball, run with other players, practice, make mistakes and learn some of the rules naturally through their own physical experience. They need to listen to a coach for direction, but then get out there with their teammates and put the game plan into action. The coach doesn't play the game, nor does he or she make the decision as to where to pass the ball once the game has started. A coach simply facilitates and lets the team collaborate and come together to figure out the best way to do things. Coaches learn through drills in practice, looking at playbooks, and, most importantly, watching videos of previous games to see what worked and what didn't. They may even observe videos of other teams and players to compare how they do things on the field. From these experiences, the coach can create an idea based on previously viewed observations and alter the practices and playbooks. So once again, the coach introduces new plays, but the players are the ones running on the field, testing these plays and working together to find the best way to implement them.

An analogy can be made with a mathematics classroom where technology was introduced to facilitate classroom dialogue—a classroom where there is not just a "coach" and a "playbook" with "players" watching a play being mapped out for them on a board. No! It is where the players actually get out there and play! This is *exactly* what technologies have brought to my mathematics classroom. Herein also lies the importance of the videos of the past classroom activities using technology and hands-on activities. I often look back to certain episodes to prepare lessons for that subject matter or simply to be prepared for potential discussions the students may bring to the experience. I almost always come across some fascinating comment, gesture or discussion that I missed in the past. Through this activity, I heighten my awareness and remind myself of my new job as facilitator and the importance of being able to adjust and take advantage of opportunities that move lessons in unanticipated, but productive, directions. These classroom episodes are proof of this dynamic and how the unexpected can be key to understanding. There is always something to be seen and it has truly taught me to never underestimate the intellectual and analytical power of my young students.

THE EPISODE

Background

The episode that I describe below was video taped in my bilingual Algebra class at a public high school in Boston. All 17 students were recent immigrants, having arrived in the United States within the past three years from Cape Verde, a group of islands off the west coast of Africa. The students were generally one or two years older than "typical" algebra students due to educational time lost during the immigration process. The students had a variety of mathematical backgrounds, but most were encountering the ideas of algebra for the first time in this course. The class was conducted mostly in Capeverdean, the students' native language, with English textbooks.

Although there are many facets to this episode that I could discuss in further detail, my focus will be on showing the interaction among the students and examining my role as facilitator of the classroom discussions. On the way, I hope to illustrate the importance of process to students' understanding and learning. Being that this is less than 5 minutes of an entire 60-minute lesson, my focus is not on what the students learned but on how the students worked together through critical thinking and discussion and made progress toward the content objective of the lesson.

The Lesson

Students were engaged in an activity in which they explored the relation between motion represented by a quadratic function and the position versus time graph of this motion (a parabola) with the aid of the Line Becomes Motion (LBM) technology. This particular LBM device consists of a pair of toy cars (similar to a die cast or larger matchbox car) that move on stationary metal tracks. This track is connected to a computer and controlled by a graphing software package. If the cars are moved by hand, the software will produce a graph that corresponds to the motion; conversely, if the user draws a graph, the software will communicate with the toy cars and they will move on the tracks according to the graph. The activity I am describing used the latter capability.

This episode is part of the first day of a series of activities on quadratic functions and parabolas. Before the students arrived, I used the LBM software to draw the graph of a quadratic function. I had the LBM set up in the front of the classroom with the computer screen covered so the students could not see the graph to which they would compare theirs. When class started, students were first asked to observe the motion of the car from their desks, which were arranged in a "U" shape around the LBM track. The students were then given an opportunity to observe the motion from a closer distance.

After observing the motion of the car a few times students returned to their seats and were asked to generate their own rough sketch of a distance vs. time graph that showed the motion of the car. When they finished, I displayed all the graphs the students had made in front of the classroom so that the students could see the various interpretations of the motion of the car. We noted three general types of graphs: Some that looked like a "V," consisting of two angled straight lines meeting at a point at the bottom; others that curved in the shape of a parabola; and one that consisted of two curved lines connected by a horizontal line at the bottom. I then invited students to discuss their observations in relation to their graphs.

Interaction Among Students

Enthusiastically the students began an open, whole-class dialogue about their graphs. Just as one student, Carol, had started to explain that she had seen the car slow down, I was called out in the hall. To my pleasant surprise, the explanations and discussion continued in my absence. The following is a segment of what occurred when I was gone:

Edgar: Sometimes, sometimes the car slows down and you think it has the same speed.

Carol: Edgar, if you look carefully over here [pointing ¾ of the way towards the end of the track] it clearly changed speed. You can see that right away.

Edgar: (*sitting at his seat away from the track*) Ok. So, Mr. Barros said that you measure from there to here [motioning from a particular point to another on the track], and see how the speed decreases, and to measure how many minutes it takes.

Carol: That is why I asked Mr. Barros for the watch. He said we don't need a watch today. Next time we can do a more accurate graph.

Edgar: So, I am asking you how you saw [that the car had changed speed]. Sometimes the car slows down or speeds up, and you think that it's going at the same speed.

Ronaldo: (*sarcastically*) Di Tera [Edgar], do you know why you could not see it? Because you do not have glasses.

Manuela: Look. If you have the same speed, it would always go like this [motions with her hand a straight line in the air], right?

Bela: Mm, hm

Manuela: But, like you are saying, if it reached here [¾ of the track] and changed speed, if it was going slower ... [inaudible] It's slower, it should go down. So, it changed direction. But, you did not show it on the graph.

Bela: Yes, it's that one [pointing at her graph, in the form of a parabola]. Look, we are saying it went back ...Do you see the second, smaller one [pointing at her graph, in the form of a parabola] over there?

When I returned to the classroom, I asked the students what I had missed. Sarcastically, but accurately, Ronaldo replied, "You missed the discussion." It wasn't until I later watched the video that I realized how right he was, and what the students were able to accomplish by discussing on their own. The videotape allowed me to see this moment that I otherwise would have missed!

The students had just finished a unit on constant speed and linear relationships. The change of speed forced some students to struggle as they adjusted their thinking and related it to the study of linear relationships. Edgar thought that before the class even began to discuss the graph, they should agree on the actual motion. He even suggested that they collect some data—perhaps time and distance data using a stopwatch, which we had done in previous classes. Carol had earlier asked me for a stopwatch and I told her that I just wanted a rough sketch of the motion, not necessarily an accurate depiction. This series of interactions is a great example of how I see students connecting one lesson to another and reinforcing their own understanding. Carol's comment, "next time we can do a more accurate graph," shows how students understood the objective of the lesson for the day and wanted to accomplish it rather than just "going along" with the class.

One of the objectives of the lesson was to establish some fundamental differences between linear and nonlinear motions. The discussion above shows how some students began dealing with the difference between speed and distance graphs generated by the car's movement. Students who felt that the car had not changed speed challenged Carol's assertion. When I returned to the classroom, Manuela reviewed for me the point under consideration and continued the discussion, "[Carol] is saying that it came to the middle of the track and it changed speed. Then, the line can't be straight. It must change, too, and it must have a bend."

In the ensuing discussion, Manuela apparently agreed with Carol that the speed changed; however, she did not think that Carol's graph was consistent with her description of the motion and argued against her own graph, which matched Carol's. This is a primary example of how students "learn to trust their own abilities to make sense of mathematics" (NCTM, 2000, p. 74). Manuela was no longer attempting to defend her graph—she had adjusted her thinking as a result of the explanation given by the other students. In the process, she had gained from the other students something she had overlooked in her original work and offered a new interpretation of the motion. I have been impressed by many such examples that I have seen of students learning through taking ownership of their ideas and expressing these ideas in dialogue with their peers.

These students felt comfortable discussing their mathematical ideas with the whole class. They were trying to understand each other's explanations, even when I wasn't there to direct them. This was a result of a classroom climate where students were free to engage in discussions, debates and be open to other's points of view. It is important to note that this kind of climate develops over time and that it takes an enormous amount of discipline on the part of the teacher to retain this dynamic. However, as can be seen here, students learn to appreciate this type of learning environment and it becomes a natural way of learning mathematics in the classroom.

While most of the students were dealing with the mathematics involved, there was at least one student, Ronaldo, who made comments that deviated from the discussion and who seemed uninterested in contributing academically. I mention this because, as a teacher, I know there is no such thing as "the perfect classroom." Although the changes I have made engage more students more of the time, there are still good and bad days and different levels of participation. As teachers we do our best to engage all of our students and to give them the best education possible by providing an effective learning environment. However, students also need to take responsibility for their own learning.

Another key point in the discussion occurred just a few lines later when the conversation shifted to the lowest point on the graph, the vertex. One student, Admar, suggested that none of the graphs, except his (two curved lines connected by a horizontal line at the bottom), showed that the car had stopped when it changed direction:

Teacher as Facilitator

Admar: When the car stops, when the car stops. You did not say that speed. You have to show that line. Also, over there [pointing at his graph, which consists of two curved lines connected by a horizontal line at the bottom] mine shows that the car stops ... the last one ... When the car stops.

Mr. Barros: Bela, Bela, not Bela. Manuela whatever you are saying over there, Manuela, I want to hear it. I want to hear it. You can't discuss just the two of you. Don't explain anything. We want to hear the things that you are saying. Save it, Manuela, Manuela, save it. I want to hear it. OK?

Mr. Barros: Admar, what were you saying? I am getting your graph, because you were referring to your graph.

Admar: I am saying that the car stopped, look. We saw that when the car went back, it stopped a little. We must represent on the graph when the car stops.

Lucindo: That is where it starts ...

Mr. Barros: Lucindo, Lucindo, let him finish and then you can take over, right? Aha. Try again.

Admar: (*gesturing with his fingers as if holding the car*) The car was going like this. As Carol said, it decreases speed. It stops a bit. Then it continued again. Now, you must represent on the graph how much the car stopped. I showed it over there [pointing at the front of the room to his graph in the shape of a parabola with a straight line at the bottom].

Mr. Barros: So, whose graph ... (*apologetically*) continue.

Admar: I showed, I represented on my graph over there [pointing to the horizontal part of his graph] how the car went like this [motions in the air]. I put a line like this where it stopped, and then another. So, the straight line represents how long the car stopped.

Student: How long the car stopped ...

Admar: Yes

Mr. Barros: So, what is wrong with the graph, either wrong or what you think that ...

Admar: I think ...

Mr. Barros: Wait, let me finish. What you think is wrong in Bela and Carol's graph ...there is another parabola here, another one there ...What can they do to improve their graph?

Admar: They must show where the car stopped. If you show the graph to someone (*Carol interjects and says "Right here" pointing to the lowest point on the graph indicating where the car stopped.*) If you show it to someone, Barros, without anyone saying anything, they would not know that the car stopped.

I became so excited about the direction of the conversation that I almost interrupted Admar's thoughts. I caught myself here, but analyzing this segment reminded me that, as the facilitator, I have to continually work to withhold my own comments until students have explained how their comments are important. Allowing students to fully explain their thinking includes stopping other students from interrupting, as I did with Lucindo. Another example of reinforcing classroom discussion norms occurred towards the end of the segment when I let Admar know that it was important to hear what I was saying as well as to give opportunity to other students to participate. This supports the idea that students need talking and thinking space to get their ideas out and that my role as facilitator is to ask questions that elevate the level of the discussion.

There are other facilitation issues in this segment that I noticed, some that I was pleased with and others not. At beginning of the segment, Admar had the floor during the whole-class discussion. As I was listening to him I noticed there was another conversation, between Bela and Manuela, that also looked very interesting. My point in addressing them was that I thought the whole class should hear that discussion, but that it was important to let Admar finish sharing his idea. Facilitating a classroom discussion can be difficult and requires constant decision making about what to dismiss because it is inconsequential to the discussion and what to highlight. The teacher is always managing multiple things while keeping the flow of the discussion moving forward.

One thing that surprised me when I watched the tape was my use of the word "wrong." I don't like my use of the word when talking about the student's work because it wasn't really about right or wrong, but instead was about what suggestion they had to better match the graph with the motion. Still I found myself using the word, and became more conscious of other messages that I might send unintentionally through my choice of words.

Admar's comment, "If you show it to someone, Barros, without anyone saying anything, they would not know that the car stopped," shows that he thinks of a graph as more than something to know to earn his grade in mathematics class. Instead, it is a powerful tool that is used to communicate important information and should be done accurately and properly. This is an example of how exposure to graphical representation using technology and motion can help students gain an understanding that a graph is a way to communicate ideas and phenomena in mathematics and real situations. Basically, that a graph is useful. To me, this is what we should strive for in education—students making the connection between real life situations and the mathematics being studied in the classroom.

The conversation continues:

Mr. Barros:	But the car stopped, didn't it?
Admar:	Nobody would know that the car stopped. If you show that graph over there (*pointing to Carol's graph, which curved in the shape of a parabola*)
Mr. Barros:	Carol has a point.
Carol:	Right here (*pointing back to the lowest point of her graph*), it did not stop. It just touched and went back. It doesn't stop...
Admar:	Can you do the car again?
Carol:	And it does a turn/loop, because over there it does not ... on this point, right here, it stop. I think that.
Mr. Barros:	Bela, oh actually, let's go to Manuela.

Mr. Barros: (*encouraging students to participate*) Do you want to tell us what you were talking about? What were you talking about over there?

Manuela: Barros, we ... (inaudible)

Bela: (*supporting her partner's observations and arguments with whom she worked together to produce a graph*) She doesn't want to say that it stops, stops. She wants to say that it changes, speed. She doesn't want to say that the car stops.

Mr. Barros: Who doesn't want to say?

Bela: Carol. We don't want to ... we ...

Mr. Barros: You and Carol?

Bela: Yeah, we don't want to say that the car stops and stays there and then goes. We want to say that it changes speed when it goes backwards and then when it goes forwards.

Admar: Barros, Barros, can you put the cars one more time please?

Mr. Barros: I'll put them on in a bit, but I want to hear more comments.

Edgar: Mr. Barros, Admar's graph shows that the car stops so many minutes. How many seconds ... 1, 2, 3 seconds. But the car doesn't stop. The car comes then it turns. Like, for example, if a car is going around a garden, it turns and takes off.

In this segment I decided to assist in the clarification of Bela's description of "stops, stops" to ensure the students' understanding of her argument. As a teacher knowledgeable of the material, it is important for me to use that knowledge to support students' responses whenever necessary, and particularly when they are likely to move the group in a productive direction. By this I mean asking questions to clarify issues that I know will help the students to better make sense of the situation and the mathematics within it.

Edgar's limited verbal participation is in only the beginning and the end of the 5-minute episode that I described, yet it was vital to the discussion. It is true that in only 5 minutes not everyone could speak, but I have found that each student who participates in class discussion adds a new dimension to the lesson. I have also observed that just because students are not talking does not mean they aren't listening. Students learn in various ways and as a teacher I need to be conscious of each of my students' learning strategies and work them into the lesson whenever possible.

Admar requested that we see the motion of the car for the second time. I find that this part of the dialogue exemplifies how the students independently want to explore their own ideas by using the technology and not necessarily turning to the teacher for the answer. To me this is crucial. To help prove his point, Admar didn't come to me to help him explain or give the answer; rather, he asked to show the movement of the car once more. This is related to Schnepp and Nemirovsky's (2001) claim that "we find students more willing to share and discuss ideas and the flawed or valid reasoning behind them when the machine is passing judgment than the teacher" (p. 95).

By sharing this episode, I wanted to show the students' dialogue and the teacher as facilitator, but I also wanted to make clear that as a result of this dialogue learning took place. Students can talk in class—agree, disagree, and have fun—but there must be movement beyond debate to connections and agreed-upon results for real learning to occur. After only 5 minutes of conversation, several critical aspects of the graph were identified which opened up more learning, dialogue and growth for the rest of the class period and the two weeks of the unit project.

CONCLUSION

My ongoing and consistent work with MCRG over the years has been an incredible experience. It has allowed me to step back and look at how I and the students, the facilitator and participants, the coach and the players, are opening up to one another, conversing and learning together. Analyzing the resulting data through reflection, watching key points of the video and engaging in enriching conversations about the students' understanding of the topic with MCRG staff has given me great insight into my teaching and my students' thinking process.

As previously noted by Schnepp and Nemirovsky (2001), I have found that my students are more willing to share and discuss and the flawed or credible reasoning behind them when the ideas are about how a device or a piece of computer software will behave, and the judgment of the ideas will come from the machine's behavior rather than the evaluation by myself as the teacher. When an idea is incompatible with the observed motion of a computer controlled car, for instance, a student must reexamine his or her own thinking to resolve the inconsistency, but does not necessarily feel judged as "wrong" by the teacher and fellow students, as all are working together to understand the device and its behavior as it is related to the mathematics being studied.

Moreover, the use of classroom technologies, such as those mentioned here, makes the classroom a fun and interesting place to learn and do mathematics, and to understand the rich connections within mathematics and its relevance to modeling everyday occurrences. When students learn to think independently and in more depth, and when they develop the ability to articulate their views in a discussion with their peers, they are building on the academic and social characters that these young adults have within but often suppress in environments that have not allowed them to express these aspects of themselves. With the changes I have made in my mathematics teaching, students are working together, posing problems, articulating, developing, deciding, debating, relating, and struggling to describe surprising results that they, themselves, have found within a lesson. I have learned, over time, to take on the role of "discussion facilitator" rather than conjecture evaluator (Schnepp & Nemirovsky, 2001) and I have found this role to be far more rewarding in the opportunity it offers to learn about my students' potential and their improvements in mathematics and communication skills.

Through my experiences as part of MCRG, I saw a whole different way of teaching mathematics, where it is not just "students look to Mr. Barros for all the answers," but instead students and teacher looking together for explanations as well as answers. Teaching mathematics well is a complex endeavor and there is no easy recipe for helping all students learn or for helping all teachers become effective (NCTM, 2000). Becoming an effective teacher requires reflection and continual efforts to seek improvement. As a part of a research group, I have had the opportunity to reflect on and refine my practices both during class and outside class, alone and with others. This kind of reflection is crucial to developing a successful learning environment for anyone at any level and has greatly improved my teaching.

The use of video is a facet of my collaboration with MCRG that is of particular interest and importance to my growth as a teacher. Watching my own classroom on video changed my way of thinking, not just in the classroom, but also even as a father listening to my young children. Video reflection has made me more aware of many things. Sitting together, the other researchers and I intensively observed the students' actions, body language/gestures, interactions, dialogue, facial expressions and the overall dynamic of the classroom. From even 5 minutes of a taped episode, we made enormous discoveries as to how the students think, feel and react to the work at hand, as well as to one another. Watching the video of a class also enabled me to recognize missed opportunities and become more skilled at taking advantage of them when they arose in my classroom. This is crucial not only for future lessons using the same devices or activities, but for following up on the "lost moments" to clarify and bring forth what students said or did during that class. As teachers we commonly plan ahead with an objective of the day, but through these videos I've realized that students are constantly bringing forth questions that go beyond the day's agenda. Pursuing these unanticipated directions is the key to empowering students to use their knowledge and critical thinking abilities to understand math. In these ways, using video has enabled me to grow into a true facilitator in my classroom where I can be most effective in helping my students.

As I engage in the challenges of teaching students now and reflect on how I might have approached them in the past, I can conclude that my collaboration with researchers, specifically MCRG, has greatly affected my ability to meet and exceed these challenges within the walls of the ever changing dynamics of my classroom and to tap into the unlimited potential that lies within each of my students. I have been able to reflect on the successes of my use of educational technologies and hands-on activities and think about how I can duplicate and exceed this success in classrooms of the future. Most importantly, the scenario I imagined and realized for my classroom and students is what I hope and encourage mathematics teaching and learning to be in classrooms and for students everywhere.

NOTES

1.   Both authors were at Jeremiah E. Burke High School at the time of the research reported in this paper.
2.   This chapter is the story of the first author. Thinking about the experience and writing the chapter was a collaborative effort of both authors.
3.   The Mathematics of Change group at TERC has included, over the time in which I have been a member, Joe Cook, Cara DeMattia, Francesca Ferrara, Teresa Lara-Meloy, Ricardo Nemirovsky, Tracy Noble, Djalita Oliveira-Ramos, Jesse Solomon and Paul Wagoner, as well as a large group of teachers who have participated over time, including Cindy Bergeron, Bob Blackler, Paul Harrison, Janice Ross, Emily Sedgwick, Antonio Centeio, Mweusi Willingham, Sydney Foster, Grace Kelemanik, Alison Kellie, Kennedy Omolo.

REFERENCES

National Council of Teachers of Mathematics. (2000). *Principles and standards for school mathematics.* Reston, VA: Author.

Schnepp, M., & Nemirovsky, R. (2001). Constructing a foundation for the Fundamental Theorem of Calculus. In A. A. Cuoco (Ed.), *The roles of representation in school mathematics* (pp. 90-102). Reston, VA: National Council of Teachers of Mathematics

CHAPTER 13

COLLABORATING TO INVESTIGATE AND IMPROVE CLASSROOM MATHEMATICS DISCOURSE

Maureen Grant
North Central High School

Rebecca McGraw
University of Arizona

My favorite thing overall is how you open each problem for discussion with the class, so we can see other ways of solving the problem and correct what we did wrong. It really irritates me when kids complain about class. Yes, it is SO easy they say. What they don't realize is that this stuff is so simple for them because somebody is teaching them well.

I really don't like the way that Mrs. Grant discusses everything. I think that she should just show how to do the work. Last year my teacher would just stand at the overhead and do examples and explain the work. I don't like the way that Mrs. Grant discusses things like we're in a history or English class.

Teachers Engaged in Research
Inquiry Into Mathematics Classrooms, Grades 9–12, pages 231–252

We didn't know whether to feel elated or to cry as we read student comments (such as those above) at the end of our first semester co-teaching a first-year algebra class. Comments such as, "I like doing group activities because if you don't understand something it is easy to have a group member explain the problem," and "I liked the bottle-filling activity because I learn better using visuals and actually doing it and interacting with it," contrasted with comments such as, "I don't like working on projects in pairs, especially when I am 'carrying someone'! It gets annoying when people ask me to give them answers," and "I don't really like the projects because we spent time on unnecessary parts. I like just getting to the main point and repeatedly working problems." These comments suggest that students' views of what occurred in our classroom—investigations of non-routine problems coupled with sharing and assessing students' varied solutions— were quite mixed. Our goals for the semester had been to put into practice the recommendations of NCTM (1989, 1991), with a particular focus on building mathematical understanding through discussion. We judged our success with respect to building discussion by the extent to which students became engaged in using discussion to develop their understandings of mathematics. We also collected data that allowed us to analyze our own teaching practices and answer questions such as: "Which aspects of teacher practice are critical with respect to building a classroom discourse community?" "What methods or strategies for facilitating discussion seem especially promising?" "What do students expect and how can we respond to their needs and expectations?" "What are the benefits and challenges of collaborating and sharing the roles of teacher and researcher?" and "What recommendations can we make to others who may want to engage in similar work?" In this chapter, we respond to these questions as we describe the genesis and development of our collaboration, the strategies we used to improve classroom discourse, and the results of our investigations—including students' feelings about "learning mathematics through discussion."

BUILDING A TEACHING AND RESEARCH COLLABORATION

Our Backgrounds

Rebecca. In my four years of high school classroom teaching, I did my best to implement NCTM's vision of teaching and learning (1989, 1991), and I felt fairly good about my classroom environment and the kinds of mathematical tasks in which my students engaged. I was never quite satisfied, however, with the quality of whole-class discussions in my own classroom. In fact, sometimes I was quite dissatisfied. I wanted students to respond to one another and I wanted their understanding of a topic to grow as they questioned and debated with one another. Too often, though, I was the one doing most of the talking, and when students talked they invariably talked to me rather than to one another.

The issue of when and how to facilitate whole-class discussions in mathematics classrooms continued to interest me after I stopped teaching high school and began pursuing a doctorate in mathematics education. I suppose it was inevitable that I should choose "facilitating whole-class discussion" as the focus of my dissertation research. Since I was no longer teaching high school, I began to consider whether one or more secondary school teachers near my university might be interested in investigating this challenging topic with me.

Shortly after choosing my dissertation topic, I began working as a research assistant on a professional development project aimed at helping experienced secondary school mathematics teachers become mentors for beginning teachers. Through this project, I was fortunate to meet teachers, such as Maureen, with perspectives on teaching and interests in classroom discourse similar to my own. When I approached Maureen about participating in a study about teachers' attempts to facilitate whole-class discussion, she agreed to collaborate with me.

Maureen. When Rebecca first approached me about collaborating, I was at a point in my professional life where I was ready to take a risk. I began my teaching career in 1979, employing a teaching style that was similar to the fairly traditional manner in which I had been taught, although with what I thought was a slightly greater concern for using hands-on activities to promote a deeper conceptual understanding of the topics being addressed. Reading the *Curriculum and Evaluation Standards for School Mathematics* [*Standards*] (National Council of Teachers of Mathematics [NCTM], 1989) prompted me to begin to examine my teaching practices a little more closely. I was excited by the vision of mathematics instruction that was represented in the lessons illustrated in the *Standards* and I began to reflect on the need to incorporate into my classroom more opportunities for students to solve problems, use their reasoning skills, and communicate their mathematical ideas.

In 1994, after 15 years of teaching experience, I began a one-year position as a mathematics consultant for a state Department of Education that included involvement with their efforts to revise the teacher licensure process. As part of the new licensure system, beginning teachers would be required to submit a portfolio consisting of videotapes and written reflections on a series of lessons. After I left the Department of Education to return to the classroom as a teacher and mathematics department chairperson, I continued my involvement with the licensure activities by receiving the training necessary to become a scorer of beginning teacher portfolios and by facilitating support seminars designed to train beginning teachers in the process of how to prepare these portfolios.

During the 1999–2000 school year, I participated in a professional development program designed to prepare experienced teachers to mentor beginning teachers in the preparation of portfolios—the project for which Rebecca was a research assistant. I became personally immersed in the process of videotaping my teaching and reflecting on my instructional practices, the same process I had been asking beginning teachers to implement in the support seminars for several years. After I was able to get past immediate reactions, such as "I say 'Okay' too much," and "I'll never wear that dress again!" I began to focus on one particular area of interest: improving the manner in which I facilitated whole-class discussion in my classroom. I had always felt that I was fairly good at talking with small groups of students about their work as they participated in hands-on activities and solved open-ended problems, but I had never felt comfortable trying to extend those conversations to the level of whole-class discussion. Rebecca's invitation provided me with the opportunity to explore my interest in a systematic way.

Setting the Stage

We[3] began our collaboration during the summer of 2000 when we met several times to consider how to teach first-year algebra in ways that would allow for more student investigation and the development of concepts as well as procedures. We realized that if we wanted students to engage in mathematical discussion, we needed to give them something to talk *about*; we needed to provide students with mathematical activities which allowed variety in approaches and solutions and which provided students with opportunities to challenge each others' thinking. We selected activities that we felt would promote algebraic thinking and classroom discourse by requiring that students look for patterns and make predictions, represent real-world situations with tables, graphs, equations and verbal descriptions, and make and test conjectures. For example, one of the selected activities required students to fill variously shaped bottles with water, one scoop at a time, graph the height of the water versus the number of scoops, and discuss how variations in their graphs related to the shapes of the bottles. In another activity, students were asked to write general formulas for geometric situations, such as finding the number of square-foot tiles that it would take to surround the perimeter of a square swimming pool. A third activity asked students to investigate functions (linear, quadratic, rational, absolute value, etc.) by comparing algebraic and graphic representations.

In addition to gathering resources and planning the scope and sequence of classroom activities, we spent time reading about and discussing strategies

for facilitating whole-class discussion. We collected information on traditional and alternative patterns of classroom discourse (Oyler, 1996; Wood,
1998), methods for establishing norms and expectations related to classroom discourse (Manouchehri & Enderson, 1999; Sherin, Mendez, &
Louis, 2000; Silver & Smith, 1996), and strategies for facilitating discussion
(Forman, Larreamendy-Joerns, Stein, & Brown, 1998; Van Zoest & Enyart,
1998). We learned much from these sources, including the importance of
(a) being explicit with students about the purposes of discussions, (b)
spending class time teaching students how to participate appropriately, (c)
attending to and consciously altering our own patterns of talk, (d) reflecting with students on the outcomes and quality of discussions, and (e) utilizing well-known, but often not well-implemented strategies, such as wait
time and the use of higher-level questions. We also learned that strategies
for altering patterns of classroom discourse can be as complex as changing
the relationships among teacher, student, and subject matter, or as simple
as waiting to respond after students speak.

We also spent quite a bit of time before the first day of class discussing
the physical set-up of the classroom, primarily the arrangement of student
desks. We decided to place the student desks in two large "U" shaped rows,
one inside the other, which extended along the sides and back of the classroom. All students would sit facing the center of the "U," which was a large
open space containing only a large table with an overhead projector.
Our hope was that the increased eye-contact between the students
would promote student-to-student discourse rather than just teacher-to-
student discourse.

We were very excited about what we were reading and discussing and
the activities we were planning for our students. We were doing more than
just thinking about how many days to spend on each section of the book
and when to give tests in order to "get through" the required material. We
were thinking deeply about the broad concepts that we wanted students to
learn and the instructional strategies that would facilitate the learning of
these concepts. This was the kind of process that each of us had wanted to
engage in ever since we read the 1989 *Standards*. We did not complete all of
our lesson-planning work prior to the beginning of the school year, but we
did begin the semester with a good idea of what had to be accomplished
and how we would go about accomplishing it. The conversations and planning sessions were invigorating and encouraging; we felt supported and
very positive about our plans for the semester.

During our summer planning we also made basic decisions about how
to gather and analyze data. Rebecca would attend two of Maureen's three
first-year algebra classes three days a week. Rebecca would videotape
the classes that she attended and Maureen would videotape the classes
on the days that Rebecca was not present. Every planning session and

every follow-up conversation would be audiotaped. Throughout the semester, we agreed to select lessons that Rebecca would transcribe from videotape, and to meet twice per month to view the videotapes and reflect on the instructional practices and classroom interactions that were taking place within them—with a particular focus on whole-class discussions. During the semester, we found audio- and video-taping whole-class discussions to be particularly useful. The audiotape was an efficient way to create transcripts of whole-class discussions, but the videotape was needed to verify who was speaking to whom. In addition, we found it useful to watch the videotape after reading and discussing a transcript. Details apparent in the videotape were sometimes absent from the transcripts, while patterns of interaction were most often uncovered by reading the transcripts.

We analyzed the transcripts by reading them individually and making notes about aspects of the discussion that seemed noteworthy; for example, instances in which students seemed highly engaged in discussion or instances in which they seemed disengaged. Because our goal was to develop our own practice as it related to building classroom discourse communities, we looked in the transcripts for methods of facilitating discussion that seemed to work or not work. We also looked for patterns in our interactions with students and considered whether the patterns were beneficial for building discussion. After reading the transcripts individually, we discussed what we had found, and watched the videotape. Finally, we discussed how the new information we had uncovered would influence our classroom instruction.

Renegotiating the Collaboration

The scenario we initially envisioned had Maureen, as the classroom teacher, doing all or most of the teaching and Rebecca, as the university researcher, observing and collecting data related to facilitating whole-class discussion. Although this might be appropriate in some situations, in our case, although we shared the planning and evaluating of instruction, the burden of implementation would have been entirely on one person. Our vision of how the semester would unfold began to change, however, on the first day of school. The first lesson began just as we had envisioned. Maureen led the discussion during the first class period and Rebecca videotaped and sat in the back taking notes. But between class periods, we decided that Rebecca would lead the discussion in the next class—and from that point on, the nature of our collaboration changed. We realized that there were valuable lessons to be learned from sitting in the back of the classroom and that we each should have a turn to do so. Likewise, we realized that we might learn more about facilitating discussion if each of us took a turn up front. So, from that point on, we became teaching and research partners and were both willing to open up our teaching practices to scrutiny and discussion. The high level of trust and mutual respect that we developed for one another—both professionally and personally—was a critical element of our collaboration.

FOCUSING ON CLASSROOM DISCOURSE

Reflecting on our efforts to build classroom discourse throughout the semester, we have identified several aspects of our practice that had a significant impact on the degree and level of discussion. Critical aspects included negotiating appropriate classroom norms and expectations, and initiating and maintaining student involvement during discussion. In the following we provide specific examples of our efforts in these areas and highlight things we learned that may be of use to others.

Negotiating Norms and Expectations

From our readings, we were aware of the importance of establishing appropriate norms and expectations for classroom discourse (Manouchehri & Enderson, 1999; Sherin et al., 2000; Silver & Smith, 1996). We felt that it was important to set the tone for the class on the very first day by creating an opportunity for students to talk with us and with one another, so we began with a discussion of the "ideal mathematics classroom" by using an introductory activity we adapted from Fleener, Dupree, and Craven (1997). We gave students several cartoons depicting various types of working environments and classrooms, such as (a) a classroom with students sitting in rows and the teacher pointing at the chalkboard, (b) a doctor consulting with a patient, (c) a group of people holding a discussion around a conference table, and (d) a laboratory with people working both individually and in groups on a variety of tasks. Students were asked to write responses to the following prompts:

1. Which cartoon shows a typical mathematics classroom? Explain your choice. Describe the roles of the teacher and the students in the cartoon you selected.

2. Which cartoon best depicts your vision of the "ideal" mathematics classroom? Explain your choice. Describe the roles of the teacher and students in the cartoon you selected.

During the whole-class discussion students shared their written responses. A few students selected the situations in which people were working in small groups or individually as "ideal" for reasons such as (a) the students in the cartoons were doing hands-on activities together rather than the teacher standing in front talking, (b) the students were being allowed to explore at their own pace, and (c) the teacher in the cartoon was interacting with either an individual student or a small group of students. The cartoon selected most frequently as both "typical" and "ideal," however, was the classroom that depicted students sitting in rows with the teacher standing in the front of the room pointing at the chalkboard. Students shared that they selected this situation because the classroom seemed well-organized, the students were attentive, and the teacher was explaining the problems step-by-step. One student did share that a potential disadvantage of this type of classroom is that some students might not "get it."

After students were given the opportunity to share their responses, we shared our vision of the ideal mathematics classroom. We felt that we needed to be explicit with students about our expectations in terms of their active participation in investigations and discussions. We explained that, in this class, students would often be asked to work in small groups and participate in discussions of their work. However, we also reassured students that there would be times in which we would be standing at the board and they would be taking notes. Looking back, we realize that this initial discussion, while important, was just the tip of the proverbial iceberg in terms of creating appropriate norms and expectations with students.

When students struggled with a mathematical task, they often looked to us to provide explicit directions that would allow them to progress through the task. We, on the other hand, felt that "wrestling" with the mathematical ideas was an important part of the learning process. Some of the tasks we assigned took two or more days to complete and many students were not accustomed to spending that amount of time working on a single problem or concept. We, however, wanted our students to learn that not all mathematical problems can be solved quickly. The following excerpt illustrates our efforts to encourage students to use each other's solutions and statements during discussion to help build their own thinking. This excerpt is from a discussion aimed at engaging students in deciding which of the equations they had previously written were valid representations of a specific linear relationship and why. The following problem had been posed in written form to the students:

Right now I am 2 miles from home. In 1 hour, I will be 5 miles from home. In 2 hours I will be 8 miles from home. Graph these points (with time from now on the horizontal axis and distance from home on the vertical axis). If the trend continues, plot a few more points that would be on this line. Write the equation of this line, define your variables.

Excerpt 1

Rebecca: These are what you came up with yesterday as possible equations for this situation: $y =$ something times x plus 2, $y = 3x + 2$, $y=3x-1$, $y = x+2$ and $y = x+3$ and we need to come to some consensus on which of these works and which of these we think don't work. (Pause). What do you think? Is there anything that you would take off that list or anything you would argue to keep on?

Brittany: . . . I don't think that the, uh, last one works.
(4 second pause)

Rebecca: Yeah, Angel and then Brittany.

Angel: I don't think that the second to last one works
(7 sec pause)

Rebecca: How come?

Angel: Because it goes by, it's three miles away from home for every hour and x was the time from home, I don't know but I don't think it's that.

Rebecca: Yeah?

Brittany: Uh, I don't think that the one above that one or the very last one. Like the $3x-1$ and the $x+3$, I don't think (it's those either)

Rebecca: Anybody help us build some arguments for any of these? . . . we have people who are doubting these last three.

Brittany: Uh, the $3x+2$ one works. (3 sec wait)

Rebecca: What makes you say that?

Brittany: Because if you substitute the x, which is the time from home, say it's at zero, you substitute the x in and it's 3 times zero is zero, and then if you add 2 on the point of the zero it's at 2, like it works, it follows the pattern.

Rebecca: Can I write it like this? (Rebecca writes $3(0) + 2 = 2$)

Rebecca: Somebody said the first one might work. . . .

Jessica: It doesn't work.

Rebecca: What doesn't?

Jessica: The first one. Because you could substitute the 3 into that but it would be the same exact one as the second one, but if you put like 4 or 5 in there it doesn't work.

Anna: Yeah it does, I tried it; I tried all the numbers.

Rebecca: Okay, so give us an example, Anna. You said I could put like 4 in there. Okay. So how did you test it?

Anna: I put it in on the graph.

Rebecca: And it made a line across.

Anna: Yeah.

Rebecca: Oh, okay. So you graphed $4x+2$ and it went across the 2, something like that (Rebecca draws a picture) and then you said you put some different numbers here and it still kept going through like this maybe (Rebecca draws some lines with different slopes going through 0,2)?

(some people start talking)

Rebecca: Yeah, speak up, what's the problem?

Brittany: I don't understand that because, yes, if you substitute x for zero, but, I don't know, if you put—

Rebecca: So if you put zero in here you get two—

Brittany: If x is like 2 hours, or if you substitute x for like 2 hours, 4 times 2 is 8 plus 2 is 10—

Rebecca: Is that right, 2 hours and 10 miles?

(a couple of no's)

Brittany: It doesn't work, and it has to work for every single equation.

We were pleased with the level of discussion found in exchanges such as the one in this excerpt. Students were sharing their reasoning and analyzing one another's solutions and ideas. At the same time, however, the students didn't necessarily see the value of the changes we had made in our classroom. Despite our discussions with students at the beginning of the year, it became clear that our vision of the "ideal mathematics classroom" was not well aligned with students' expectations. After a few weeks, when we still hadn't passed out textbooks, some students began to complain, asking, "When are we going to start doing algebra?" One 9th grade student expressed concern because we were not on Chapter 2 of the textbook like all the other algebra classes, and she worried that she was going to be unprepared for the state-required high school Graduation Qualifying Exam (GQE) that she would take in the fall of her 10th grade year.

Students who had been successful in more traditional classroom settings continued, throughout the semester, to express frustration with the time we spent on discussion. We realized that if we wanted students to feel some ownership of the classroom and if we wanted them to continue to participate in classroom discussions, then we needed to respect their opinions and address their concerns. To do this, we consciously incorporated problems from the book into our lessons and periodically talked with students about why we were asking them to investigate and discuss and what they could learn from doing so. While we continued to push students to think, we also attempted to provide additional scaffolding for students who were struggling. Some of the strategies we used included (a) suggesting "paths" to students that we knew would be productive, (b) asking them draw a picture or graph, and (c) requesting that they write down three things they had discovered about the problem and one or two questions they wanted to ask. Overall, we found that supporting students and respecting their desires while still engaging them in reasoning, problem solving, and communicating was more of a dilemma to be managed than a problem to be solved. We also recognized that our initial norm-setting was insufficient.

In addition to finding ways to better incorporate the textbook into our teaching, scaffolding student thinking, and being explicit about our reasons for using discussion, we have identified an additional strategy that might have proved useful in focusing students' attention on the purpose and value of discussion. Although *we* spent time thinking about the purpose and value of whole-class discussions, we did not often engage *students* in reflecting on the influence of particular discussions on their own learning. In addition, only about half the students in each class regularly spoke during discussions. Other students were very quiet and we knew little about the extent to which they were actively listening during discussions and, if they were listening, how hearing the ideas of other students was or was not helping them. One way we could have engaged students in thinking about the value of discussions and also gained greater insight into students' thinking would have been to incorporate post-discussion writing into our regular classroom activities. Asking students to write about their role in a discussion and what, if anything, they learned during the discussion would have encouraged meta-cognition, communicated to students that we value their opinions about discussion, and helped us to understand the ways quieter students experience and hopefully benefit from whole-class discussions.

Initiating and Maintaining Whole-class Discussion

It feels like discussion is a top or something, something you have to get spinning, something you have to do something to [in order] to get it moving. And once it's actually moving, it's OK. Then you just have to keep it going.

When we reflected on our efforts to facilitate whole-class discussion, we often found ourselves making comparisons such as the one above, made by Rebecca. Our efforts to motivate a need for discussion by choosing appropriate tasks, structuring the space in the classroom in order to promote student-to-student discourse as well as student-to-teacher discourse, and attempting to establish norms and expectations for the classroom environment all seemed to fall into the category of *initiating student engagement in discussion*. Other aspects of initiating student engagement in discussion included consideration of how and when to incorporate discussion into the classroom, allowing students sufficient time to engage in tasks so that they could develop opinions that were worthy of discussion, presenting students' ideas to the class, and motivating listening. But once a discussion was initiated, we were concerned with keeping students engaged in the conversation—we wanted to "keep the top spinning." We wanted students to talk with one another and ask one another questions instead of just talking to us. We wanted them to agree and to disagree and to try to understand and build upon one another's ideas. We wanted them to do more than just "share" their ideas—we wanted them to "discuss" their ideas.

One questioning strategy that we used to initiate discussion was gathering students' solutions and re-presenting them to the class for everyone to consider (as in Excerpt 1). This strategy seemed to work particularly well when different answers, or different representations of the same answer (e. g., $2x + 2 = y$ and $2(x + 1) = y$) were possible because students then had to think about whether an answer different from their own might also be correct. Another strategy we used for getting discussion started involved introducing a new problem for the class to consider that was related to, but perhaps more complex than, the ones they had been working on in small groups. The small group work prior to discussion was still very important, because students needed to have developed some ideas of their own in order to feel that they had something to say to the whole class. But because the problem was a new and complex one, everyone had something to think about (i.e., no one was likely to feel that they knew everything about the solution) and the problem was likely to require the combined effort of all to solve, as in Excerpt 2.

Excerpt 2 is taken from a whole-class discussion in which students were given a set of 64 linear and nonlinear equations. Most of the nonlinear equations were quadratic. Students were asked to think about what the graph of each equation would look like and then group equations together that they expected to have similar graphs. Next, students graphed the equations using graphing calculators and reanalyzed their groupings. We began this project after students had spent several weeks writing linear equations to describe relationships between two quantities given a description of a relationship (as in Excerpt 1), a data table, a graph, or some combination of the three. The excerpt that follows was taken from a discussion that occurred after students had created their initial groupings but before they began graphing. The excerpt illustrates strategies we used to encourage discussion and shows how students responded to our efforts.

Excerpt 2

Rebecca: I'm going to put an equation up here that's not on your list but you're going to recognize; it has some similarities to what's on your list.

$$y = \frac{3(x)^2 - 9}{3}$$

Rebecca: So my first question to all of you is, what are some characteristics of that equation?

Student: It has exponents.

Rebecca: Mindy?

Mindy: The variables it uses, x and y, are the same as the ones on our sheet.

Rebecca: (gestures to Anna)

Anna: It has fractions, you divide.

Rebecca: Who can add to what Anna said about the fraction issue?

Jessica: There's a division bar all the way across.

Rebecca: Cathy, did you want to say something? . . .

Cathy: The parentheses.

Rebecca: Who can build off Cathy's idea and say something more about parentheses?

Jessica: There's a variable in it.

Rebecca: Who can say some more about that?

Robin: There's an exponent outside the parentheses.

Rebecca: Anything else we can say about parentheses in this situation?

Jessica: There's not a problem in it.

Rebecca: Who thinks they can explain what Jessica means? I see a few hands.

Anna: There's no multiplication or adding in there, just a variable. . . .

Rebecca: Let's move on to a different question . . . . How about this?

Rebecca writes:

$$y = \frac{3(x)^2}{3} \quad \text{[Equation A]}$$

Rebecca: What do you think about graphing the first equation and then this one? I know they look different. Any thoughts? (5-second wait time) I'll throw up another one and see if that sends your mind working.

[Rebecca writes two equations on the overhead projector.]

$$y = \frac{3(x) - 9}{3} \quad \text{[Equation B]}$$

$$y = 3(x)^2 - 9 \quad \text{[Equation C]}$$

Brittany: All you're really doing is taking a step out. Like in [Equation A] you're just taking out the step of subtracting 9 and then on [Equation B] you're taking out the step of the exponent and on [Equation C] you're taking out the step of dividing by 3.

Rebecca: So which of those bottom three do you think might look a lot like the top one? (8 seconds wait time) Adam?

Adam: [Equation C] because all you are doing is not dividing by 3, so it will be parallel to the first one, the line will.

Pat: I think [Equation A] might look the most like it, because it's like, at high numbers multiplication would affect the line of the graph more and minus 9 at a high number like 300 and something wouldn't affect the line as much.

Brittany: Well, what if it was a low number?

Pat: Then it would. I'm just saying.

Rebecca: Anyone want to respond to Adam or Pat or throw in their own idea? (10-second wait time)

Rebecca: What I want to say to everyone to focus you [as you start] graphing is, what we are going to try to do is get at this kind of "family of equations" issue. We started out with this one [Rebecca points at the original equation] and I made some [other equations] that are a little bit different, but are kind of related, have some similarities. Maybe this one [Rebecca points at Equation A] belongs in a family with this one [Rebecca points at the original equation], or this one [Rebecca points at Equation B], or two of them or all three of them. And so a big piece of this project is getting some ideas in your own minds about equations and their graphs. . .

[At this point students move into their small groups to begin the graphing portion of the project.]

Once the discussion top had started spinning, we continued to encourage students to build upon and respond to the comments of previous students by stepping back from the center of the discussion both verbally and physically. Verbally, not speaking after each student's comment was a method we used to break the teacher–student–teacher–student pattern of talk. This sometimes caused lengthy pauses in conversation, as seen in Excerpt 2. It took some time for us, and for students, to become comfortable with the silence, but this strategy seemed to be particularly useful. In addition, we frequently reminded students to "talk to the class" as they spoke. With respect to stepping back physically, moving to the side of the classroom was helpful in terms of de-centering ourselves and creating space for students to speak directly to one another; we also found *not* looking at students as they spoke was helpful in terms of encouraging the speaker to look at other students and remove the focus from the teacher.

Analyzing discussion transcripts helped us to differentiate among patterns of interaction in our own whole-class talk (before the semester began we had read about patterns other researchers had identified). We noticed when student-to-student talk occurred and began to develop strategies to try to foster it, such as explicitly asking one student to respond to another student (as in Excerpt 2: "Anyone want to respond to Adam or Pat . . ."). We also examined the extent to which we were successful in utilizing the discourse-enhancing strategies we read about, such as wait time and high-level questioning, and we tried to understand what caused us to regress occasionally into more traditional, teacher-centered speech patterns. In our case, it seemed that regressing to traditional speech patterns mainly occurred when students were not fully engaged in discussion (e.g., when there was "off-task" noise or inattention to the problem at hand) or when students' preparation for discussion was not sufficient and they were therefore unable to analyze the mathematics to the extent we'd hoped for. When students were "off-task" we found ourselves tempted to "pull" them in by speaking loudly and dominating the room with our voices and body positions. When students were not able to engage with the mathematics in the ways we wished, we tended to do more telling.

When we analyzed the discussion of which Excerpt 2 was a part, we noticed a marked difference between the first part of the discussion, in which students list off characteristics of equations, and the second part, in which students consider the relationships among several equations. In the first part, a variety of students were ready to participate, but the driving question was a relatively low-level one, and the pattern of talk was consistently teacher–student–teacher–student–teacher–student. Rebecca's effort to engage students in building upon each other's ideas by asking "Who can add to what Anna said about the fraction issue?" may serve to keep students listening, but does not serve to raise the level of discussion much. Engaging the class in developing a list of characteristics of equations may be a reasonable way to encourage students to think about equations in terms of their component parts, but we would not consider an exchange such as this an example of whole-class discussion as there is no evidence of debate or student–to-student talk.

The question Rebecca asked, "Which of those bottom three do you think might look a lot like the top [original] one?" is an example of putting forth a more complex problem that is related to students' previous small group activities. In this case, the problem was more complex because the equations combine an exponent, multiple operations, and a pair of unnecessary parentheses. A relatively long wait time was needed to give students time to think about this new problem, but a few interesting hypotheses were generated almost immediately. Students also began to respond to each other, although a long wait time after Brittany, Pat, and Adam's comments did not generate more participation. At this point, Rebecca decided to cut short the dialogue because of time constraints. Had we decided to devote more time to pursuing students' ideas about this problem, we would probably have given them some individual "think time" or perhaps a few minutes to discuss their ideas about these equations in small groups—two strategies we used frequently to prepare students to participate in discussions.

We can also see how our efforts to initiate and facilitate discussion played out in the classroom in Excerpt 1. To begin this discussion, we had collected students' responses to the problem and re-presented them to the class (written on the overhead projector) for analysis. Rebecca attempted

to share responsibility with students for assessing the validity of the equations when she said, "We need to come to some consensus," "Anybody help
us build some arguments for any of these?" and "Is there anything that you
would take off that list or anything you would argue to keep on?" Pausing
after students spoke (and looking around the room during the pause) was
a strategy we used to encourage other students to jump into the conversation rather than waiting for one of us to speak. If no one spoke after several
seconds, we typically asked a "why" or "how did you know" question of the
previous speaker (Rebecca did this in Excerpt 1), or directed a question to
the entire class such as "Who can build on what _____ said?" Sometimes
a statement such as, "Somebody said the first one might work," proved useful for encouraging student engagement.

In summary, we found that successful facilitation of whole-class discussion depended on both allowing sufficient time for students to develop
and/or reflect upon their own thinking prior to discussion and using strategies such as wait time and high-level questioning. In terms of maintaining
discussion, physically removing ourselves from the center of the room was
critical so that students could speak directly to one another. As previously
stated, we also found that continued attention to establishing norms and
expectations regarding the role of discussion in learning mathematics was
essential to maintaining student engagement.

Working Together to Investigate and Improve Our Practice

Throughout the semester, we were co-planners, co-teachers and co-"reflective practitioners." Some days one of us would teach one period and the
other one would teach the other period. Since the person who taught during the second of our two class periods together had a slight "teaching
advantage" gained by watching the first person stumble through uncharted
territory, we took turns being the lead teacher of the first class period.
Even when one of us was the primary teacher of the lesson, the observer
would always participate by working with students when they were involved
in small group activities and discussions, and would sometimes interject
ideas into the lesson from the back of the room. For one lesson, we divided
the class into two groups and each of us taught.

Although we saw ourselves as equal partners in the collaboration effort
because we were co-planning and co-teaching, there were several factors
that caused students to view our roles in the classroom as somewhat
unequal. One student wrote at the end of the semester, "Another thing I
like is that you have two teachers even though one is not the real teacher."
Of course, the teacher who had been hired to teach the classes by the
school district (Maureen) was still in charge of assigning grades, contacting
parents, and serving as the primary disciplinarian in major discipline

issues. The teacher/researcher (Rebecca), on the other hand, was not in the classroom every day, and we tried to implement the more interesting activities (i.e., implementing non-routine problems, student investigations, and projects) on the days when she was there. Tests, quizzes and other more "routine" activities were usually saved for the days when only Maureen was in the classroom. Despite these differences in our roles, some students did view us as a team, making comments such as, "I think it is a good idea how both of you teachers go around and check everyone's work so you would have a better understanding on where everyone's weaknesses and strengths are," and "You two are too strict," while other students clearly had a preference for one teacher or the other.

Taking time out of our busy days to reflect on the process of facilitating whole-class discussions was critical to the development of our facilitating abilities. We met at the end of each week to have an extended discussion about the week's activities and to plan for the next week, but we also consulted with each other before, during, between, and after class on a daily basis. Questions we asked each other included: (a) Do you have any last-minute ideas about how we should refine this lesson? (b) What solutions or solution methods are the students in your small groups coming up with for this problem? (c) How should we proceed with this lesson based upon what we have observed in the small groups? (d) How can we improve this lesson for the second class period? and (e) What changes or adjustments should we make to tomorrow's lesson in light of students' reactions and understandings of today's lesson?

When we reviewed segments of our instruction (transcripts or videotape), we asked questions such as, "What could the teacher have done differently here?" "What question could have been asked next that would have prompted more discussion from students?" "Should the teacher have interjected her own thoughts at this point in the discussion or should she have waited longer for students to respond to one another?" "Would it have been better for the teacher to be sitting down or standing to the side of the classroom during this part of the discussion to indicate that she wanted to facilitate and participate in the conversation rather than dominate it?" We became aware of some of our habits, such as repeating what students had said rather than requiring them to listen carefully to one another and to speak loudly enough so that other students could hear. We were excited by moments in the lessons when students engaged in extended dialogue with one another without our interjections; we discussed our thoughts about why this had happened and what we could do to ensure that this would happen more often during classroom discussions. We also discussed the sometimes painful internal dialogues that we had with ourselves as lessons unfolded and we struggled with issues such as allowing students to form their own conceptual understandings versus "telling" and wondering whether we should answer questions that students had not asked. For teachers like us, who were normally isolated in our own classrooms, these conversations were a powerful form of professional development as we had the opportunity to discuss what was happening with another professional who shared similar goals and attitudes.

CONCLUSION

We began our semester together eager to find ways to better facilitate whole-class discussion; by the end of the semester, we felt a little bit uncertain about how successful we had been. While we were able to engage students in the types of tasks and activities advocated by the NCTM *Standards* documents (1989, 2000), we knew that students had mixed reactions to our efforts. Our observations of student work and various other methods of assessment (e.g., written work, presentations, tests and quizzes) suggested that students were meeting our goals for their mathematics learning, but not all students had come to value investigation and discussion as a method of instruction. In addition, there remained aspects of the process of facilitating discussion that concern us. We were only moderately successful in encouraging student-to-student talk and the development of mathematical arguments during discussion. Some students didn't talk much, and we are still unsure about the appropriate role of quiet and/or shy students during whole-class discussion.

Although we were not entirely successful with respect to all aspects of engaging students in using and valuing discussion as a means of learning mathematics, we have learned a great deal—both about facilitating discussion and about collaborating to improve practice. Specifically, we have learned that negotiating norms and expectations is an ongoing process that may require as much attention in October and November as it does at the beginning of the school year. Students' expectations are an important part of the classroom culture and must be addressed—in our case, this meant incorporating the textbook more frequently and finding ways to scaffold student thinking. In addition, if we were to repeat the semester, we would also engage students in reflecting on the influence of particular discussions on their own thinking. With respect to engaging students in discussion, we found that allowing students sufficient time to engage with tasks and develop ideas and opinions prior to discussion was especially important, as was the use of well-established strategies such as wait-time and high-level questions. Re-presenting students' ideas to the class and introducing a new but related problem were also helpful in terms of encouraging discussion, as was verbally and physically de-centering ourselves.

With respect to collaborating to improve practice, we found it beneficial to intermingle the roles of teacher and researcher. We discovered that working closely with a like-minded colleague was a powerful professional development experience. Insights that might have remained nascent took form as we discussed each day's events and planned for the weeks ahead. Watching videotapes and reading and discussing transcripts helped us focus further on the particular aspect of our practice that was of most interest to us (i.e. facilitating discussions). Working together, each of us learned more than either of us could have alone. We know that partnerships among teachers and researchers take many fruitful forms; however, for us, it was the blending of our roles that made our experience such a powerful one.

NOTE

1. A third collaborator, a middle school mathematics teacher, was also part of the study and joined us in our early planning discussions and in periodic debriefings throughout the study.

REFERENCES

Fleener, M. J., Dupree, D. N., & L. D. Craven, L. D. (1997). Exploring and changing visions of mathematics teaching and learning: What do students think? *Mathematics Teaching in the Middle School 3*(1), 40-43.

Forman, E., Larreamendy-Joerns, J., Stein, M., & Brown, C. (1998). "You're going to want to find out which and prove it": Collective argumentation in a mathematics classroom. *Learning and Instruction 8*, 527-548.

Manouchehri, A., & Enderson, M. C. (1999). Promoting mathematical discourse: Learning from classroom examples. *Mathematics Teaching in the Middle School 4*, 216-222.

National Council of Teachers of Mathematics. (1989). *Curriculum and evaluation standards for school mathematics*. Reston, VA: Author.

National Council of Teachers of Mathematics. (1991). *Professional standards for teaching mathematics*. Reston, VA: Author.

National Council of Teachers of Mathematics. (2000). *Principles and standards for school mathematics*. Reston, VA: Author.

Oyler, C. (1996). *Making room for students: Sharing authority in room 104*. New York: Teachers College Press.

Sherin, M., Mendez, E., & Louis, D. (2000). Talking about math talk. In M. Burke & F. Curcio (Eds.), *Learning mathematics for a new century* (pp. 188-196). Reston, VA: National Council of Teachers of Mathematics.

Silver, E. A., & Smith, M. S. (1996). Building discourse communities in mathematics classrooms: A worthwhile but challenging journey. In P. C. Elliott & M. J. Kenney (Eds.), *Communication in mathematics, K–12 and beyond* (pp. 20-28). Reston, VA: National Council of Teachers of Mathematics.

Van Zoest, L. R., & Enyart, A. (1998). Discourse, of course: Encouraging genuine mathematical conversations. *Mathematics Teaching in the Middle School, 4,* 150-157.

Wood, T. (1998). Alternative patterns of communication in mathematics classes: Funneling or focusing? In H. Steinberg (Ed.), *Language and communication in the mathematics classroom* (pp. 167-178). Reston, VA: National Council of Teachers of Mathematics.

CHAPTER 14

PROFESSIONAL DEVELOPMENT AS A CATALYST FOR CLASSROOM CHANGE

Michael Verkaik
Holland Christian High School

Beth Ritsema
Western Michigan University

The story I[1] am about to tell began at a high school mathematics department meeting in the fall of my 13th year of teaching. Our school had agreed to participate in a National Science Foundation grant called Renewing Mathematics Teaching through Curriculum (RMTC). RMTC's goal was to assist teachers in implementing mathematics programs based on the National Council of Teachers of Mathematics (NCTM) *Standards* (NCTM, 1989, 1991, 1995, 2000). Our district had already adopted Connected Mathematics Project and Core-Plus Mathematics Project (CPMP) curriculum materials and was interested in the professional development assistance provided by RMTC. To participate in RMTC, our district needed a representative for the RMTC Collaborative Council. Since no one volunteered, I

Teachers Engaged in Research
Inquiry Into Mathematics Classrooms, Grades 9–12, pages 253–272
Copyright © 2006 by Information Age Publishing
All rights of reproduction in any form reserved.

offered to represent our school. Little did I know how that decision would drastically affect my classroom teaching! This chapter tells how investigating my teaching over a four-year time frame has molded my thinking and consequently my actions as a classroom teacher.

My story is simple. I was an experienced mathematics teacher and felt very comfortable in the classroom setting. I had good control; the students were respectful and quiet when I was talking. Most students were learning and I assumed that the ones who were not learning either were not trying hard enough or simply lacked the ability to fully understand the mathematics. Most of my students were successful at solving problems that required straightforward mathematical calculations. Word problems, however, were a different story. Not many students were able to figure them out. But, if I repeatedly showed the method and changed only the numbers on the test, students were moderately successful. I wanted my students to do well and not to struggle and I thought the best way to meet that goal was to do more examples and assign more problems. My approach seemed to enable most students to complete the problems. Yet two questions began to surface regularly in my mind: (a) How are students thinking about mathematics? and (b) Why do many students have difficulty applying their knowledge to new problems, even problems that are only slightly different?

Participating in the RMTC project provided me the opportunity to think about and discuss teaching and learning with colleagues. I began by participating in two week-long summer workshops organized around teaching the mathematical content of the Core-Plus Mathematics Project courses. The student investigations in this problem-based curriculum exposed me to a new way of developing mathematical concepts in my classroom. Following the workshops, I reorganized my classroom to allow student collaboration on the investigations in the curriculum. By listening to student conversation and interacting with groups that were struggling, I had a window into student thinking and was fascinated by what I heard. I wondered whether my failure to listen carefully to student thinking in the past had caused me to miss opportunities to help students learn. I became much more focused on the question: "How can I assist my students in *their own thinking?*"

Perhaps my interest in improving my teaching is what spurred me to volunteer to be videotaped during the RMTC Collaborative Council meeting. As a group of educators, we had decided that a valuable way to improve our teaching would be to look more closely at episodes of classroom teaching. Specific teaching episodes would allow us to base our discussions on concrete teacher and student actions. Video would allow us to put the teaching into slow motion and dissect the thinking and questioning that occurred in the classroom.

Two weeks later, an investigation from the Core-Plus Mathematics curriculum on linear equations and inequalities titled "Equivalent Rules and Equations" was videotaped in my ninth-grade class. Typical mathematics lessons in my class are broken down into three parts: a whole class launch, a student group investigation, and a whole class closure summarizing the learning. The tape was reduced to key teaching episodes during the lesson and a transcript was created of those episodes to allow for more precision in the analysis. The transcript and tape were then viewed and discussed by a group of 25 middle and high school teachers. The principal investigators for the RMTC Project facilitated this Reflecting on Teaching session. (See Grant, Kline, & Van Zoest, 2001, for details on the session format.) What a learning tool! Imagine being able to back up a conversation, slow it down, and think about what the other person was saying. This allowed me to begin to understand what students were thinking. In addition, examining the videotape allowed me to look carefully at my teaching decisions and their effect on student thinking. I need not rely on a fleeting memory of the class period. Having colleagues discuss the tape with me provided insights that I would not have gained from watching the tape alone.

The Reflecting on Teaching session focusing on video clips of my classroom lasted 5 hours. A long list of issues emerged:

- What makes a lesson launch effective?
- What is the teacher's role when students are collaborating on an investigation?
- What is happening in the classroom when students are learning?
- How do you help students understand problem contexts?
- What makes a question "good"?
- Some teacher decisions are made in launches, investigations, and closure with the goal of using time effectively. How do these decisions affect student learning?
- How does altering the curriculum materials by providing more structured handouts affect learning?
- How does teacher perception of the mathematical goals of the lesson influence teaching decisions?
- How do we identify productive and nonproductive student struggle?
- What makes lesson closure effective?

In the following, I describe four areas where watching and analyzing the videotape with colleagues led to significant changes in my teaching: effective lesson launches, problem contexts in learning, effective teacher questioning, and students' struggles while learning mathematics.

INVESTIGATION 4   *Equivalent Rules and Equations**

From network television, movies, and concert tours to local school plays and musical shows, entertainment is a big business in the United States. Each live or recorded performance is prepared with weeks, months, or even years of creative work and business planning.

For example, a reasonable cost for production of a CD by a popular recording artist is $100,000. Depending on various things, each copy of the CD could cost about $1.50 for materials and reproduction. Royalties to the composers, producers, and performers could be about $2.25 for each CD. The record label might charge about $5 per copy to the stores that will sell the CD. Using these numbers, how does the *label's net profit P* relate to the *number of copies N* that are made and sold?

1. Here are four possible rules relating *P* and *N*.

   $P = 5.00N - 1.50N - 2.25N - 100,000$

   $P = 5.00N - 3.75N - 100,000$

   $P = 1.25N - 100,000$

   $P = -100,000 + 1.25N$

   a. How was each rule constructed from the information given?

   b. Which of the four equations seems the best way to express the relation between copies sold and profits? Explain your reasoning in making a choice.

   c. Compare tables of (*sales, profit*) data to see the pattern defined by each of the four equations. (Consider CD sales of 0 to 500,000 in steps of 10,000.)

   d. Create graphs of the four rules with your calculator or computer. Compare the graphs to see the pattern each gives relating sales and profits. Use the trace function to examine each graph.

* Reprinted with permission from *Contemporary Mathematics in Context,* Course 1, Unit 3 © 2003 published by Glencoe/McGraw-Hill.

Figure 14.1.   Activity 1

EFFECTIVE LESSON LAUNCHES

The lesson that was videotaped provides an example of a lesson launch in my classroom when I first began teaching the Core-Plus Mathematics program. The investigation in the lesson, from the third unit in Course 1 (ninth-grade mathematics), focuses on algebraically equivalent expressions. The first problem (shown in Figure 14.1) directs students to compare four equivalent expressions for calculating profit for a musical production company. On first encountering them, students are not likely to see all four expressions as algebraically equivalent. In fact, they might initially think that only the equation expressing profit (P) in terms of the number of CDs sold (N) accurately represented the situation.

In planning the launch of this investigation, I considered three options: (a) telling students the expressions were equivalent, (b) having them try to figure it out themselves in small groups, or (c) leading them through steps that would make the equivalence clear. I rejected option (a) because I did not want to take away their discovery of this fact. I rejected option (b) because I felt that they had too little background to recognize the equivalence without my help. As a result, I decided to lead the whole class through the symbol manipulation needed to recognize equivalence. I thought that this approach would have the benefits of refreshing their memories of pre-algebra skills and providing experience with symbolic work to make the remainder of the investigation go more smoothly.

Watching this launch on videotape helped me realize that the whole class discussion was a lengthy, tedious process. Much of the time was teacher-centered dialogue with one or two students. It took over 20 minutes of valuable class time to achieve my goal. The following is a representative interchange from the launch.

Teacher: Look at the equation $P = 5N - 1.5N - 2.25N - 100,000$. What does the P represent?

Joe: Profit.

Teacher: What are they subtracting all the expenses from?

Joe: What you get from selling the CD.

Teacher: Which we call...?

Joe: Income.

Teacher: Good! Income. So, where is the income in this problem?

Joe: N

Teacher: N is the number of copies of CD's

Joe: The $100,000...

Sally: P

Teacher: P is our profit.

After watching this lesson on videotape, it became clear that my approach in the launch had resulted in detracting from the main emphasis of the investigation—identifying equivalent equations. Furthermore, watching the students struggle with the symbolic work during the group investigations provided evidence that the twenty minutes spent in the whole class explanation had little, if any, positive residual effect for the students.

When we viewed the videotape, my colleagues and I were able to offer options for launching the lesson that I hadn't considered in my initial planning—options that didn't require students to rely on algebraic manipulation skills they didn't all have. For example, students might check equivalence by looking at graphs or tables of the four equations. Once the four expressions for profit were identified as equivalent using these mathematical representations, students could have gone back to examine the problem setting and the symbolic forms to understand the equivalence. This alternative approach would have helped focus students on the mathematical goal of the lesson and also think about the mathematics for themselves.

HELPING STUDENTS WITH THE PROBLEM CONTEXT

In preparing for this same lesson, I expected that students would have difficulty with the context of the problem. Income, expenses, and profit are not familiar concepts for most ninth-grade students. Thus, I had an example in a different context ready to help students understand the problem in the student text.

Already during the launch students were expressing confusion about the use of the terms "profit" and "income." As you can see in transcript in the previous section, students were essentially guessing at a response to my question: "So, where is the income in this problem?" Since this line of questioning did not seem to be generating understanding of the difference between profit and income for Sally or Joe, I introduced a new context even before setting students to work on the investigation.

Teacher: Profit is different than income. Income is the amount of money I take in. For example, if I sell books and sell you a book for $12.95. Do I make $12.95?

Sally: No.

Teacher: What cost me something?

Sally: The book?

Teacher: The book. I had to pay for the book. Now, maybe I only paid $4.00 for the book, so my profit is $12.95 – $4.00, so

> where is my income in this problem? (Pointing to the text-
> book problem projected from the overhead)

Joe: $5.00

Teacher: $5.00 for...

Joe: Every CD I sell

Assuming that the class now understood the difference between "income" and "profit," I let them begin the investigation. Unfortunately, there was evidence in the tape that most students were still not clear about the difference between the two terms.

There was a second time that I veered from the context in the text. After assisting a few groups attempting to find a shorter form for the equation $P = I - (20 + 0.25I)$, I went to the board and did a class demonstration. (See Figure 2 for Activity 2 Part d, the problem they were working on.) I began by telling students that since you can not combine the 20 and the $0.25I$, you have to subtract them separately: $I - 20 - 0.25I$. Then, I introduced another context that I had prepared ahead of time to help students understand the distributive property. I said,

> Suppose that I had a twenty dollar bill and I owed Heather $5 and Ben $10. I could say I owe $15 because I can combine them together and subtract it from $20 and have $5 left. However, if I could not combine them I could go to Heather and give her $5 and then go to Ben and give him $10, leaving me with $5. I could subtract them together, but also separately.

Thinking that this explanation using money would help students simplify $I - (20 + 0.25I)$, I set them back to work on their investigation. It wasn't long before the following exchange with one of the groups occurred.

Teacher: I is my income

Denise: Yes. And you have this 0.25 income too.

Teacher: Correct. So, if I was trying to subtract this quantity I can subtract the whole thing at once or I can subtract them...

Denise: Separately.

Teacher: Separately.

Jordon: You would do I minus $0.25\,I$; then plus 20

Teacher: I have to subtract both of them though, right?

Jordon: Well...

Teacher: I have to subtract them separately.

Jordon: I minus 20 and then I minus 0.25 I

Teacher: So, first you would minus 20

Denise: That would be two equations

Teacher: Ryan, how would you write it?

Ryan: I minus 20 then I minus 0.25I

Teacher: If I do it like that, then I am subtracting it from I each time. I want to subtract the 20 and then the 0.25 I.

This exchange showed me that my explanation of the distributive property in a new context did not help build understanding. Rather, it confused students. Notice that Jordan's and Ryan's understanding of "subtract them separately" meant to make two separate expressions, $(I - 20)$ and $(I - 0.25I)$. Instead of using the context of the income problem to understand the distributive property, they were trying to use the method I had presented. Analyzing the tape made it clear that the way I directed students during the exchange encouraged them to follow the method I had presented.

At the end of the class period, I was helping one of my brighter students, James, and I asked him, "How do you handle a minus sign in front of the parenthesis?" James replied, "I don't know; I forget." I did not even catch it at the time, but when I watched the tape I realized that my thoughtfully crafted launch and example in a new context had done nothing to help even this student understand how to simplify the expression.

Examining this record of practice has led me to think differently about introducing alternative contexts during a lesson. I now think more carefully about how to help students navigate the context in the text, since that will be the context needed to understand and apply the mathematical goals of the lesson. I rarely introduce a new context unless it seems to provide access to the current context for some students who may not have the necessary prior knowledge. If I do introduce a new context, I watch for clues indicating whether or not students have transferred the mathematics learned from the new context to the current problem in order to assess the effectiveness of my decision.

The complexity of teaching decisions has become more apparent to me. Although I have analyzed the interactions above on the basis of the problem context, there may be other variables influencing the effectiveness of my instruction. For example, the words I used to facilitate the new context may have affected its usefulness. In the next section I look more carefully at how I use words.

EFFECTIVE TEACHER QUESTIONING

By watching myself on tape and listening to the discussion, I learned that often I listen to students with a hidden agenda. I wanted to help students find the correct answer. My approach was usually to provide hints and fill-in-the-blank type statements until the problem was answered to my satisfaction. As we discussed the videotape, I noticed that many of my comments and questions were directing students to the answer rather than helping me understand their thinking or encouraging their thinking. For instance, the following conversation with Katie is about eliminating parentheses in order to find a shorter form for the equation $P = I - (20 + 0.25I)$.

Teacher:	What do you know about parentheses?
Katie:	You can like add all these together.
Teacher:	But you can't do that, right?
Katie:	Ya.

What else would Katie say? It would have been much more appropriate for me to ask, "How can you do that?" Then, I would have been able to determine if she knew that she could not combine terms. I was focused on getting her to the answer (usually by my preferred teaching method) rather than trying to assess her thinking.

Studying the transcript of the videotape following the Reflecting on Teaching session, I was drawn back to the exchange following my explanation of subtracting $5 and $10 from $20. This exchange is repeated below. Focus this time, as I did, on the way that I directed the discussion. Recall that these students were working to come up with two other equations for profit as a function of income equivalent to $P = I - (20 + 0.25I)$ [Activity 2 Part d, in Figure 14.2].

Teacher:	I is my income
Denise:	Yes. And you have this 0.25 income too.
Teacher:	Correct. So, if I was trying to subtract this quantity I can subtract the whole thing at once or I can subtract them...
Denise:	Separately.
Teacher:	Separately.
Jordon:	You would do I minus 0.25 I; then plus 20
Teacher:	I have to subtract both of them though, right?
Jordon:	Well...
Teacher:	I have to subtract them separately.
Jordon:	I minus 20 and then I minus 0.25 I
Teacher:	So, first you would minus 20
Denise:	That would be two equations
Teacher:	Ryan, how would you write it?
Ryan:	I minus 20 then I minus $0.25I$

INVESTIGATION 4   *Equivalent Rules and Equations**

2. Studios that make motion pictures deal with many of the same kinds
   of cost and income questions as music producers. Contracts sometimes designate parts of the income from a movie to the writers,
   directors, and actors. Suppose that for one film those payments are:

   > 4% to the writer of the screenplay;
   > 6% to the director;
   > 15% to the leading actors.

   a. Suppose the studio receives income of $50 million from the film.
      What payments will go to the writer, the director, the actors, and
      to all these people combined?

   b. Let I represent film income and E represent expenses for the writer,
      director, and actors. Write two equivalent equations showing how E is
      a function of I. Make one of those equations in a form that shows the
      breakdown to each person or group. Make the other the shortest
      form that shows the combined payments.

   c. A movie studio might have other expenses, before any income
      occurs. For example, there will be costs for shooting and editing
      the film. Suppose those pre-release expenses are $20 million.
      Assuming there are no other expenses, what will the studio's
      profit be if income is $50 million? (Remember, the profit is
      income minus expenses.)

   d. Here is one expression for profit P as a function of
      income I for that movie.

      $$P = I - (20 + 0.25I)$$

      - In your group, discuss how this rule was constructed from
        the given information.

      - Write two other equivalent expressions for profit as a function
        of income. Make one of those equations as short as possible.
        Make the other a longer expression that shows how the sepa-
        rate payments to the writer, the director, and the actors affect
        profit.

   * Reprinted with permission from *Contemporary Mathematics in Context*,
   Course 1, Unit 3 © 2003 published by Glencoe/McGraw-Hill.

Figure 14.2.   Activity 2

Teacher: If I do it like that, then I am subtracting it from I each time. I want to subtract the 20 and then the $0.25\ I$.

Why didn't I ask any questions? In retrospect, questions I might have asked following Jordan's first comment are: "What do the I and the $0.25I$ mean in this problem context?" "How do these pieces contribute to finding the profit?" "What does the expression $20 + 0.25I$ represent?" "Why is that expression inside parentheses?" Questions like these would have focused students on the meaning of the symbols and helped me know whether or not they were using the context of the problem to reason about the symbolic expressions.

When I analyzed the transcript to see what had motivated me to introduce the example of subtracting \$5 and \$10, I traced it back to this exchange:

Teacher: What do you know about parentheses?

Eric: That you have to do them first.

Teacher: Ok, is there anyway that we can get rid of the parentheses?

Eric: You can add all these together.

My response to Eric was to tell him that he could *not* just add them all together. In this exchange I did ask two questions. These questions elicited responses about what Eric would do, but not why he thought he could add all of the terms together. I wonder what direction the discussion and the students' thinking would have taken if I had let the group try their ideas. I might have proceeded in this manner: What is the result when you add them together? OK, you now have two ways to express the profit for the record label. How can you verify that they are equivalent? Discussions with my colleagues helped me recognize that students might have tried graphing the two expressions simultaneously. Students would have noticed that the expressions were not equivalent. I should have encouraged students to confront this inconsistency instead of trying to get them to think about the problem the way I was thinking about it. Questions like: "Does that make sense?" (both when it should and when it shouldn't) meet more than one of my goals as a teacher. This type of question helps me understand student thinking; it sets the expectation that mathematics should be reasonable or explainable; and it can be used to help students learn one way to monitor their own problem solving.

Going into the lesson, I had been focused on developing students' understanding of how to distribute the negative sign over parentheses. When the problem asked them to write two other expressions for profit as a function of income, I wanted students to manipulate the current equation. My questioning was leading students to do that, however, it was not making sense to them. Why didn't I ask students how the rule $P = I - (20 + 0.25I)$ was constructed from the context information as was done in the textbook? (See Figure 14.2 Part d, first bullet.) Did students skip this discussion? Did they superficially discuss it? I don't know because I didn't ask them. I could have sent them back to discuss that item.

Besides thinking about my own questions while teaching, examining the video has alerted me to missed opportunities to utilize the textbook questions that have been thoughtfully designed to help students develop understanding of the mathematics concepts. I now recognize that sometimes my teaching moves subvert the intent of the investigation. My preparation for teaching now involves more careful examination of the questions in the student text to inform my thinking.

STUDENT STRUGGLE

Discussion around these classroom exchanges provoked my thinking about student struggle during the launch and group investigations. I had thought my role was to make the path easy for the students—to get out all the bumps in the road where they might have difficulties. I had thought my role was to prevent student struggle, but I now began to wonder if I should be orchestrating struggle. Was a certain level of confusion needed to ensure learning? Were my attempts to make the path easy for students, in effect, reducing opportunities for students to think and thus, to learn?

Student struggle is a bi-product of teaching using problem solving. Managing the struggle becomes one of the greatest challenges of the classroom. I now believe that when teachers carefully guide students to correct answers, they are denying students an opportunity to make sense of the mathematics being studied. As a result, their ability to retain and use the new mathematics concepts and skills will be limited.

I have witnessed some wonderful exchanges in the classroom when students engaged in thinking about a particular task that they found difficult were, through productive struggle, able to make sense of it. However, too much struggle can lead to lack of production, frustration, and discipline problems. A big challenge of classroom management is sorting out productive struggle from unproductive struggle and learning when to step into the discussion. My students and I are beginning to talk about this issue. We are learning that times of unproductive struggle are times to stop, think, and ask questions like: "What are we missing here?" "Should we try a different approach?" "Why do we seem to be stuck? What does this question mean?"

It is also important to determine why students are struggling. This is another benefit of video taping your classroom. By looking at the discussions that were taking place in my classroom, I became aware of how sometimes teachers can be part of the reason students are struggling. For example, when the students were asked to write an equivalent expression for $P = I - (20 + 0.25I)$, they were struggling to find one. I spent a significant amount of time trying to get them to change that expression symbolically—without showing them how—and we were spinning our wheels. If I had allowed them to come up with different expressions from the context, where one of them is $P = 0.75I - 20$, then there might have been productive struggle in which deep learning could take place as they saw how the equations they created could be manipulated into different forms. A bi-product would have been the learning of the distributive property. The strategies they developed could then be used to solve new problems.

Student struggle always needs to be monitored to ensure that learning takes place. If a person is a new mechanic, her supervisor does not want her to struggle opening the hood of every car. Struggling with fixing the engine is where the focus needs to be. That may seem obvious, but after studying this clip I realized that I was doing that with my students. They were struggling with a mathematical situation, but in the wrong place. The following interchange among a group of students determining if four profit equations are indeed equivalent illustrates the point of orchestrating struggle:

Teacher: The first question says here are four possible rules. Are those rules correct?

Marie: No

Derek: Yeah

Sonya: No

Teacher: Well maybe we should start there and decide whether or not they are correct.

Marie: Do we just say incorrect?

Teacher: Well

Marie: For that, there is with a dollar twenty-five so.

Teacher: Read number 1

Derek: Here are four possible rules for P.

Teacher: So, here are four possible rules, do they leave room for you to doubt whether the rules are correct?

Marie: No

Derek: It's possible.

Sonya: Yeah

Teacher: So what does possible mean?

Marie: That it could be right.

> Sonya: That it could be wrong.
>
> Teacher: That it could be wrong? Maybe we'd better start with that. Look at the first rule.

Notice how this group seems to be spinning their wheels. They disagree about whether the rules *are* equivalent and I focus their attention on the wording of the question. I want them to be thinking about why these four rules are equivalent and discover properties that are commonly used in mathematics, but ended up focusing them on getting the hood open. Instead of letting the floundering continue, I could have asked them how they could determine if the expressions are equivalent. The students themselves may have suggested looking at the tables or graphs. Then, they could have seen that the expressions were in fact equivalent. By doing this I would have focused students' energies on recognizing the symbolic equivalence of the expressions.

The video was a tremendous asset in determining how I was orchestrating student investigation work in the classroom. Before videotaping my class, I remember sometimes leaving a class surprised by the particular topics students had difficulty understanding. I thought about student struggle, realized that there were class periods where students struggled with topics that I did not expect them to, but had no idea how to address this issue. I never thought about the difference between productive and unproductive struggle nor about preparing questions prior to class to promote student thinking. By examining the teaching episode and re-thinking what had happened, I began to better anticipate areas of potential struggle and prepare more helpful questions prior to the instructional time.

CONTINUING TO IMPROVE MY TEACHING

Armed with new ways to think about teaching, I continued to look for an answer to my question, "How can I help my students learn better?" My class preparation changed. I recognized that thinking about what students might struggle with prior to a lesson is beneficial, but not for the purpose of making the road easier. It is useful to formulate questions to focus the struggle on the desired lesson objectives. My questions should help me assess how students are thinking about the problem and then focus students on the intended outcome. Then, I can at least try to make connections between their thinking and the problem at hand. I now recognize that providing students *my* solution path can be confusing for them as they take their thoughts and try to put them in my mental schema. Two articles that have been particularly beneficial in my thinking about questioning are *Questioning Your Way to the Standards* (Mewborn & Huberty, 1999) and

Unveiling Student Understanding: The Role of Questioning in Instruction (Manouchehri & Lapp, 2003).

 As I think about problem contexts in investigations, instead of having another example ready for students, I now have questions prepared to help me understand what they are thinking and to push their thinking about the problem. Questions such as "Why did you decide to start that way?" and "What do you mean by that statement?" help me understand student thinking. In addition, these types of questions focus the student on his or her work instead of following my steps. My goal is to find a window into student thinking. Looking back at the exchanges has helped me to guard against explaining my thinking to students by asking students to complete my sentences or asking questions that direct students to the answer.

UNDER THE MICROSCOPE AGAIN

In the year following the Reflecting on Teaching session based on my classroom, I became a teacher leader in the RMTC project. I participated in institutes to develop my ability to co-facilitate Reflecting on Teaching sessions with my co-author, Beth. As the two of us planned sessions based on video excerpts of other teachers' classrooms, delivered the sessions for colleagues, and debriefed the sessions, we continued to discuss the craft of teaching. I took our ideas and tested them in my classroom. Even though RMTC had formally ended, when Beth proposed that we videotape my class investigating the same lesson ("Equivalent Rules and Equations") two years after the initial videotaping, I eagerly agreed. As we discussed the new tape, it allowed me to observe improvements and to identify areas I wished to work on.

 One thing we noticed is that my goal for the launch had changed. I set the stage for the investigation rather than trying to help students avoid struggle. I did not go through the first problem with them this time; instead I let them wrestle with the idea of equivalence in their groups. This gave all students time to think about equivalence instead of my dialoging with one or two students prior to setting students to work.

 Overall, I was satisfied with the new launch. It seemed to meet my goal for setting the stage, and it was a much better use of class time. However, I still needed to answer the question: "What can I do to assist students who do not see the equivalence?" I was disappointed to notice that even though I mentioned using a variety of methods in the launch (graphs, tables, and symbol manipulation), I seemed to be unaware of those methods when I was involved in discussions with students. The first two groups I worked with both had members that thought the four expressions were not equivalent. I spent a significant amount of time trying to help them see that they were. I used better questioning techniques, but as I watched the tape I realize that I never asked them, "How do you know they are not equivalent?" "How can you tell?" These questions might have focused them on tools they had available, such as graphing the expressions as functions. If students had first recognized that the expressions were equivalent, they could have studied the expressions to identify the equivalence symbolically. I again missed an opportunity to encourage productive thinking.

Another area of growth was my ability to involve all students in small group discussions. The following is an excerpt from a discussion during my second taped session when the students were struggling with the question: "What will the studio's profit be if income is $50 million?" (Activity 2 Part c, in Figure 14.2). Notice how the dialog is mainly between the students as they struggle to make sense of the situation.

Michael: The producer's profit, it is not all those combined to equal? But you gotta take the 20 out of the 50 and then take the 6% from there.

Nikki: You do take this combined and minus this from... you minus 20 from 50 and minus the combined from 50 and you get your answer

Tony: That's how you do it, right? [looking at teacher]

Teacher: Am I your answer book?

Tony: Well, we know we're right, he just thinks...

Nikki: No

Teacher: Well

Nikki: He thinks he just has to take this 4%. I don't know why he is doing that?

Teacher: 4% of what?

Tony: No, 6%

Teacher: What do you take 6% of?

Tony: The producer. The producer's profit. Because it is not asking for the actress or the screenplay writer

Michael: No, but he has to pay these people. He has to take this much out of how much the profit is because he has to pay these people for what they are doing. This isn't the producer's

Nikki: But the producer doesn't get the left over though does he?

Tony: No that's how much money the movie's making ... profit.

In the first taping the conversations followed a much different pattern of teacher-student-teacher-student. Generally it was the same student and myself with the other group members remaining relatively quiet. Careful study of the first tape helped me realize that I needed to get other group members involved in the conversation. In the second tape I asked questions such as: "How about you Emily?" and "Are you all in agreement that the first one is correct?" These questions are now a natural part of my facilitating. I have found that as I involve all of the members of the group in the discussion students rely on each other more, resulting in deeper, more productive group discussions.

An analysis of my second tape turned up additional evidence that student discussion and thinking was enhanced when my conversation included statements and questions such as the following:

- Talk about which one of the rules you think is not correct.
- Well, it says, "Here are four possible rules." So, that's something you guys have to hash out as a group. Are they all right?
- So, convince Emily that that is true.
- Why? Where did those numbers come from?
- Why is that one wrong?

When I did not lead the students with questions focused at getting to the answer, there was a dramatic change in the discourse among students. In the following segment the students are trying to figure out where the $1.25 came from in the fourth rule (Activity 2 Part a, in Figure 14.1).

Teacher: So where did that $1.25 come from?

Steve: They simplified the expenses on the cost per CD.

Becky: That is not right.

Teacher: When you say simplified expenses...

Steve: That means the expenses for the $5.00 per copy sent to the stores are $2.25 in royalties and the copies for $1.50 are simplified into $1.25.

Becky: Well it can't be. Because ...

Trina: He's not taking away the same amount.

Becky: Yeah, then you are only taking away $1.25 instead of ...

Steve: You are not taking away the $5.00 ...

More of the students in the group were involved in the conversation as they contradicted each other and tried to make sense of each other's thinking. Students were required to think hard about their approach to the problem as they articulated their strategy to solve the problem. Focusing on the students' thinking also allowed me to listen and identify incorrect reasoning. For example, two of these students thought that the $5.00 was an expense, not income.

CONCLUDING REMARKS

My experiences searching for ways to improve my teaching have changed my thinking about student learning and my role in this process. Often students can learn when they are told what to do, but I now recognize that they can learn more if they are given the opportunity to think. I have learned that there can be many ways to think through a problem. When I was telling students my solution path, I was stifling their thinking. By pushing students to use my method, I was encouraging them to adopt a particular procedure that they may or may not have understood. I am convinced that students learn best when they are given time to discuss their approaches. Doing this gives students the opportunity to make connections and gives me an opportunity to assess their thinking. My role is to ask questions that help me assess students' thinking, help them focus on the problem, and teach them how recognize unproductive struggle and reorient themselves to a new approach.

My classroom is quite different today than earlier in my career. My students spend much of their time trying to figure out problems and mathematical properties. I enjoy listening to my students use the language of mathematics as they communicate with each other. Without thoughtful listening and asking questions that probe their thinking, I cannot assist students in developing deep understanding. Teaching is a complex venture. It seems that the more I learn, the more I realize what it is I still need to learn. The support and pushing provided by discussing actual teaching episodes with friends and colleagues advanced my quest to improve learning opportunities for my students. And so, equipped with greater knowledge and improved skills, I continue on the journey to improve my teaching and my students' learning.

ACKNOWLEDGMENTS

This chapter is based in part on work supported by the National Science Foundation under grant ESI-9618896. Any opinions, findings, and conclusions or recommendations expressed in this article are those of the authors and do not necessarily reflect the views of the National Science Foundation.

NOTE

1. Although this chapter is written in the voice of the first author, the process described here was a collaborative journey. Initial collegial support within the RMTC professional development activities was followed by a second round of videotaping and ongoing conversations between both authors that led to the collaborative writing of this paper.

REFERENCES

Grant, T., Kline, K., & Van Zoest, L. R. (2001). Supporting teacher change: Professional development that promotes thoughtful reflection on teaching. *Journal of Mathematics Education Leadership, 5*(1), 29-37.

Manouchehri, A., & Lapp, D. (2003). Unveiling student understanding: The role of questioning in instruction. *Mathematics Teacher, 96*(8) 562-566.

Mewborn, D., & Huberty, P. (1999). Questioning your way to the Standards. *Teaching Children Mathematics, 6*(4), 226-227, 243-246.

National Council of Teachers of Mathematics. (1989). *Curriculum and evaluation standards for school mathematics.* Reston, VA: Author.

National Council of Teachers of Mathematics. (1991). *Professional standards for teaching mathematics.* Reston, VA: Author.

National Council of Teachers of Mathematics. (1995). *Assessment standards for school mathematics.* Reston, VA: Author.

National Council of Teachers of Mathematics. (2000). *Principles and standards for school mathematics.* Reston, VA: Author.